I. Bejtka

Cross (CLT) and diagonal (DLT) laminated timber as innovative material for beam elements

Titelbild: Cross laminated timber beam – finite element model and
normal stresses distribution

Band 17 der Reihe
Karlsruher Berichte zum Ingenieurholzbau

Herausgeber
Karlsruher Institut für Technologie (KIT)
Lehrstuhl für Ingenieurholzbau und Baukonstruktionen
Univ.-Prof. Dr.-Ing. H. J. Blaß

Cross (CLT) and diagonal (DLT) laminated timber as innovative material for beam elements

This work was supported by a fellowship within the Postdoc-Programme of the German Academic Exchange Service (DAAD) during my stay at the Department of Wood Science, University of British Columbia, Vancouver/Canada from September 2007 to August 2008.

by
I. Bejtka
Karlsruhe Institute of Technology (KIT)
Timber Structures and Building Construction

with a total thickness of 500 mm. The maximum plate dimensions a limited to 30 m in length and 4.8 m in width. These maximum dimensions are only limited by the dimensions of the press and manufacturing tools. Fig. 1 shows the typical assembly of a CLT element with five layers. Three layers are orientated in the main axis direction while two inner layers are orientated perpendicular to the main axis direction.

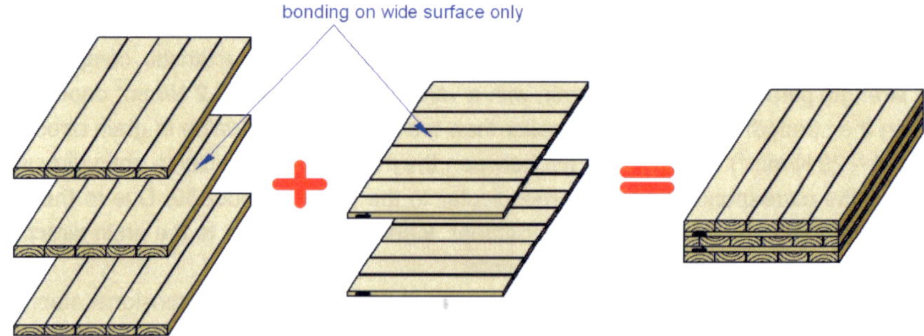

Fig. 1   From raw material to cross laminated timber (CLT) panel

In Fig. 2 a typical CLT panel manufactured by a European manufacturer is displayed. As displayed right in Fig. 2 the layers are bonded together only over the wide surfaces. The gaps between neighboured boards are used to prevent stresses between the crosswise orientated layers due to disabled shrinkage or swelling. Additional grooves (right in Fig. 2) can even reduce these stresses, in particular in thicker CLT elements. Furthermore, the missing of bonding between neighboured boards reduces the production effort significantly.

Fig. 2   CLT panel with gaps and grooves

CLT panels with large dimensions provide the opportunity to use this massive wood product even in large buildings. In Europe, multi-storey cellular structures are typical applications for CLT products where CLT can be used for wall, roof and floor elements. An example for a building which was made solely using CLT panels is displayed in Fig. 3. To connect the elements together, traditional dowel-type fasteners (nails, bolts, dowels and screws) were used. Hence tension, compression as well as bending can be transferred between the CLT elements.

Fig. 3   Massive timber building made using CLT panels

A further advantage of CLT panels is the prefabrication. Openings for windows and doors as well as the final shape of a wall or floor system can be prefabricated in a factory. Just the final assembling is to be done on construction site. Compared to systems, which are produced on construction site, prefabricated elements are more accurate and economical.

The dimensional stability is one more advantage of CLT elements compared to solid wood. Due to the crosswise orientation of the raw material, CLT shrinks and swells equal in the main axis direction and perpendicular to the main axis direction. According to [1], the swelling and shrinkage parameter for CLT panels is 0.02% per length unit and per 1% moisture changing. Hence, the dimensionally stability of CLT is 12 times higher than those of solid wood or glulam perpendicular to the grain direction. For this reason, solid wood and glulam are prone to splitting while CLT is not. Compared to solid wood and glulam with orthotropic strength and stiffness properties, the strength and stiffness values for CLT panels are nearly similar in the main axis direction and perpendicular to the main axis direction.

Currently, CLT elements are commonly used in the massive wood building systems as elements loaded in the out-of-plane and in the in-plane direction. Due to the previously mentioned advantages, CLT could be used as well for beams loaded in bending. In particular, beams loaded by tensile stresses perpendicular to grain or by high shear stresses are prone to splitting. Reinforcing using timber screws, glued-in rods, etc. is only efficient in local areas. For beams, where tensile stresses perpendicular to grain or/and high shear stresses occur along the biggest part of the beam (see Fig. 4), traditional reinforcing methods are inefficient.

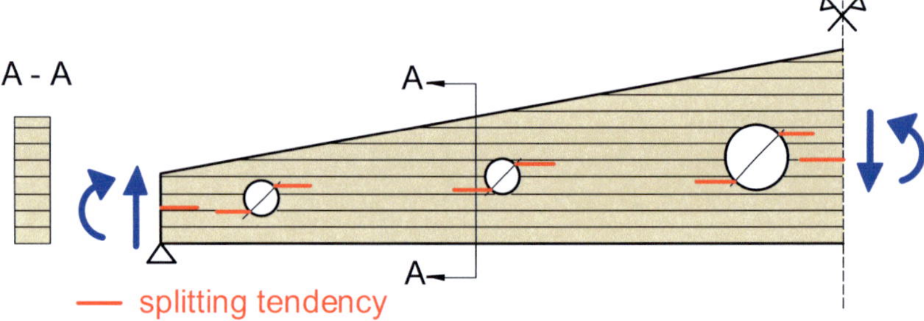

Fig. 4   Glulam beam loaded by tensile stresses perpendicular to the grain and by shear stresses

For those problematic beams, which are prone to splitting, CLT instead of solid wood or glulam could be used .Therefore CLT must be orientated on edge (Fig. 5). In this case, axial loads are transferred by the parallel orientated layers, while the perpendicular orientated layers are used to transfer loads perpendicular to the beam axis direction. Hence, additional reinforcements wouldn't be necessary any more. Using CLT elements instead of solid wood and glulem would increase the dimensional stability, too.

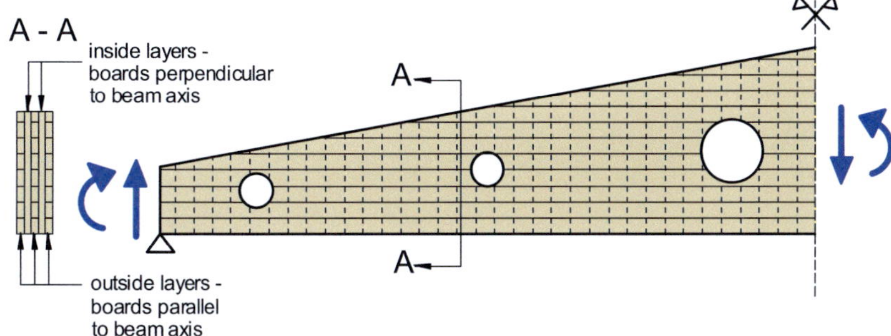

Fig. 5   Beam loaded by tensile stresses perpendicular to the grain made using CLT elements

Within the scope of the DAAD founded research project, the possible field of application for CLT beams was determined. It is shown, that CLT provides an opportunity for beams loaded in bending and by tensile stresses perpendicular to the grain direction. Due to some disadvantages, the field of application for CLT in beams is strongly reduced. For this reason, a new product, diagonal orientated timber (DLT), was developed. Tests were done on self-made CLT and DLT beams. The test results confirm the accuracy of the finite element analysis, which was done to investigate the load-displacement behaviour of CLT and DLT beams.

The investigations were performed at the Department of Wood Science, University of British Columbia in Vancouver/Canada and kindly supported by Prof. Frank Lam.

2 Preliminary considerations

Compared to solid wood and glulam, CLT used in beams loaded in bending is characterised by one significant disadvantage: Due to the crosswise orientation of the layers, the bending stiffness of CLT beams is usually smaller than the bending stiffness of solid wood or glulam beams with identical dimensions. Particularly for wide spanned beams, the performance of beams loaded in bending is controlled by the deflections and hence by the bending stiffness. Therefore, for long beams CLT is not suitable.

Theoretical, the bending stiffness of CLT beams is controlled only by the parallel to the beam axis orientated layers. At best, the bending stiffness of CLT beams can be calculated considering the parallel orientated layers as a full shear plane. In fact, even the parallel orientated layers cannot be assumed as a full shear plane because the neighboured boards in the same layer are not glued together. Therefore, traditional CLT as used for panels may not be usable for beams loaded in bending.

To counterbalance this disadvantage of CLT, following modifications could be done:

- The bending stiffness of edgewise orientated CLT elements loaded in bending increases with increasing ration between the parallel and the perpendicular to the beam axis orientated layers. However, with increasing ratio between the parallel and the perpendicular to the beam axis orientated layers the reinforcement performance degreases. The thinner the perpendicular orientated layers are the lower loads perpendicular to the beam axis can be transferred.
- The bending stiffness of edgewise orientated CLT elements loaded in bending could be increased by additional gluing of boards along the narrow surfaces. Then the parallel orientated layers could be considered as a full shear plane.
- Another opportunity to increase the bending stiffness of edgewise orientated CLT beams loaded in bending could be given by optimizing of the reinforcing layers. Usually, CLT is made with crosswise orientated layers, where the layers are orientated perpendicular to each other. By an orientation of the inner layers in diagonal direction, like displayed in Fig. 6, the bending stiffness could be increased.

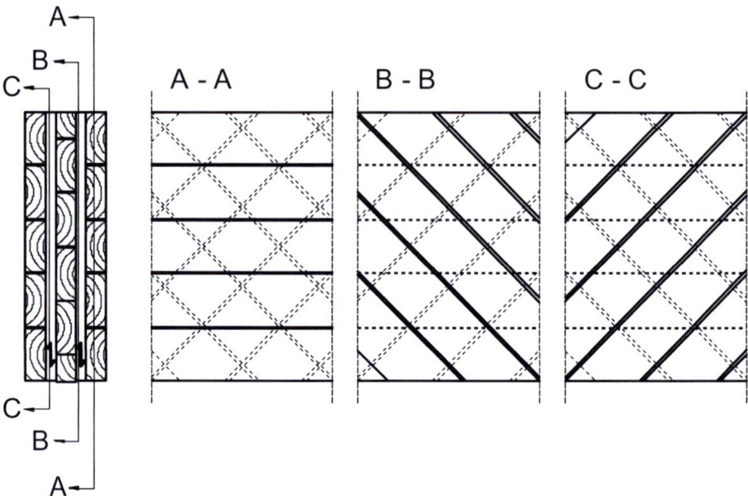

Fig. 6   Modified CLT element with diagonal orientated inner layers

To estimate the relationship between the local modulus of elasticity in the beam axis direction and the inclination angle, numerical calculations were done. Therefore, a beam with two planes loaded in bending was investigated, where the planes were orientated in the opposite direction to each other (Fig. 8). Both planes were glued together using contact elements and both planes considered the orthotropic material properties for the timber. The modulus of elasticity in grain direction was assumed as $E_0 = 12800$ N/mm^2 while the corresponding value perpendicular to the grain direction was assumed as $E_{90} = 275$ N/mm^2. The angle between the beam axis and the grain direction for the front side plane is α. In contrast, the inclination for the back side plane is $(180° - \alpha)$.

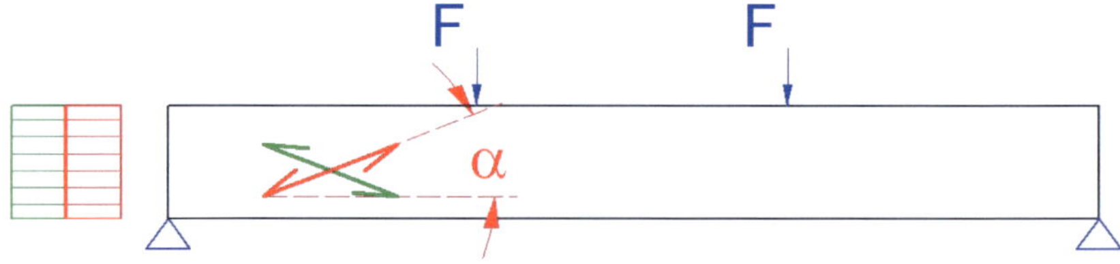

Fig. 7   Model to determine the relationship between the local MOE and inclination

The relationship between the local MOE in the beam axis direction and the inclination α is displayed in Fig. 8.

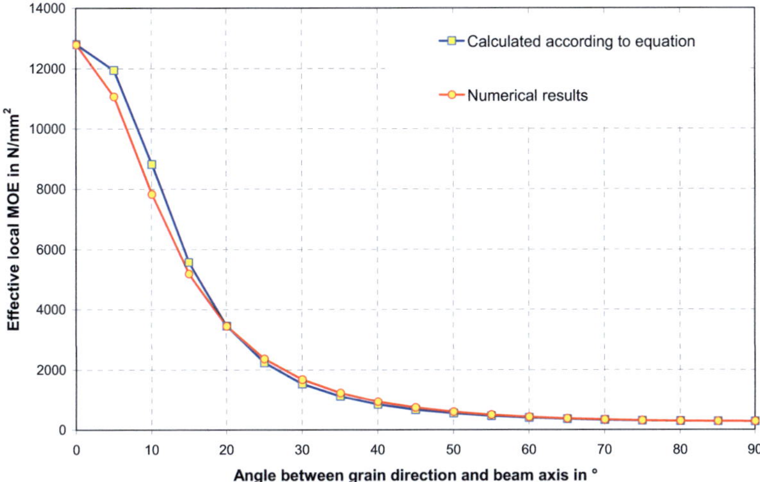

Fig. 8   Relationship between inclination and local MOE in the beam axis direction

This relationship can be described as well using equation   (1). This equation fits the numerical results very good for n = 2.6.

$$E_{ef} = \frac{E_0}{\dfrac{E_0}{E_{90}} \cdot \sin^n \alpha + \cos^n \alpha} \tag{1}$$

According to the results (Fig. 8 and equation   (1)) the local MOE degrease very fast with increasing inclination. Already at an inclination of α = 13°, the local MOE in the beam axis direction degreases to the half of the MOE in grain direction. For an inclination of α = 45° the local MOE in the beam axis direction can be calculated to $E_{ef,45°}$ = 663 N/mm². This value is about 2.4 times higher than the assumed E_{90} = 275 N/mm² but unfortunately much smaller than the assumed MOE in grain direction (E_0 = 12800 N/mm²).

This result shows that even diagonal (α = 45°) orientated reinforcing layers are not able to increase the bending stiffness of a CLT beam loaded in bending significantly. Nevertheless, this new product with diagonal orientated layers (DLT) can provide other opportunities: CLT with perpendicular orientated layers is suitable for beams which are loaded both in bending and by tensile stresses perpendicular to the beam axis. The modified product with diagonal orientated reinforcing layers (DLT) could be suitable for beams which are loaded both by tensile stresses perpendicular to the beam axis and by shear stresses simultaneously. High tensile stresses perpendicular to the grain direction and high shear stresses occur for example in local areas of beams. Therefore, the modified product with diagonal reinforcing layers was investigated as well within this project.

3 Practical part of the work

3.1 Preliminary remark

To determine the performance of cross (CLT) and diagonal (DLT) laminated timber elements used in beams loaded in bending tests on short and long beams were done. Short beams were planned to evoke shear failure. Against it, long beams were used to evoke bending failure. Tests were done on common beams with rectangular cross-section and without any singularities. To determine the reinforcing effect, additional tests on beams with singularities like holes and notched supports were done. Finally, the tests were done taking into account different cross-section types. Therefore, CLT and DLT beams with five layers and DLT with four layers were tested.

In the framework of this research project, all specimens were produced in the former "CANFOR" research lab in New Westminster, British Columbia where a heated press is located. The press is able to glue panels with 4 m in the length, 2 m in the width with a total thickness of 14 cm. The press is equipped with a heater and with vertical and horizontal hydraulic jacks. Unfortunately, the hydraulic jacks, which were orientated perpendicular to the plane surface, are limited to a total load of 850 kN. Considering the maximum press dimensions (4×2 m^2) and the maximum loading (850 kN), a total pressure of 0.11 N/mm^2 can be applied on the elements.

Most of the glue manufacturers recommend a pressure of about 1 N/mm^2 to obtain an adequate bonding between the timber members. With lower pressure the bonding performance degreases. Due to the limited pressure, it was not able to produce beams with large dimensions.

3.2 Timber selection

A CLT panel is a high quality product which is usually made using low quality raw material. Even low damaged raw material can be used for CLT. Damages, like smaller cracks or bigger knots in the boards don't affect the load-carrying capacity of CLT elements significantly. For this reason, not damaged raw material but as well raw material which was previously damaged by the mountain pine beetle (MPB) were used for the specimens.

Following material was provided by the Department of Wood Science for the test specimens:

Series 1: **1512 boards with a length of 244 cm** **H x B = 89 mm x 38 mm**
Series 2: **756 boards with a length of 244 cm** **H x B = 94 mm x 21 mm**

The standardised "2-by-4" boards with a total length of 8' (244 cm) were 1.5" (38 mm) thick and 3.5" (89 mm) height. This raw material (series 1) was taken from woods, where the mostly trees were destroyed by the mountain pine beetle (MPB). 756 pieces were taken from trees, which were cut down after a period of three years by being affected by the mountain pine beetle. The remaining 756 pieces were taken from trees which were cut down after a period of one year by being affected by the mountain pine beetle. Boards from series 2 were not attached by the mountain pine beetle.

For all 2268 boards the timber density, the moisture content and the dynamic modulus of elasticity were determined. Each timber member was numbered and hence all determined parameters could be used to allocate the properties of CLT and DLT beams. Finally, the results were used as well to determine the relationship between the damage due to the MPB and the stiffness properties.

The frequency distribution for the dry timber density is displayed in Fig. 9. The dry timber density for timber damaged by the MPB and for the non-attacked timber is quite similar. The frequency distribution for the moisture content considering all 2268 boards is displayed in Fig. 10. The moisture content for timber which was affected by the MPB for three years is smaller than the moisture content for timber which was affected by the mountain pine beetle for one year. Thereby, both timber packages were stored for around one year at the University of British Columbia. Timber from series 2 was previously dries in an oven.

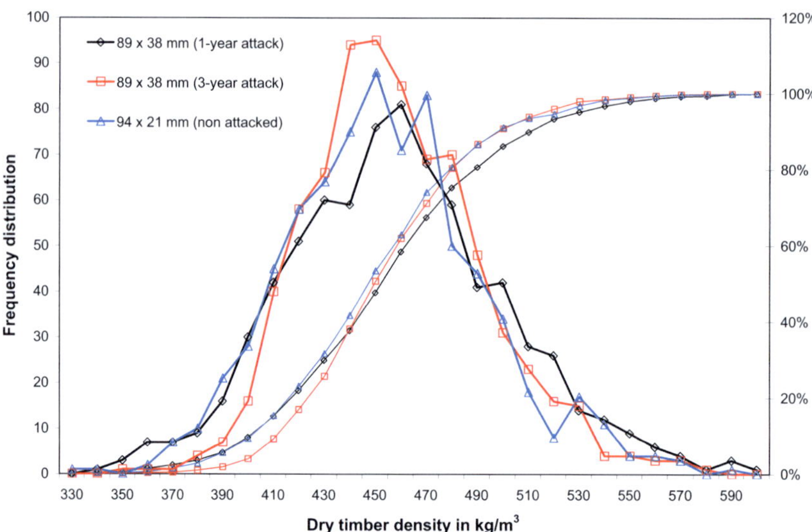

Fig. 9   Frequency distribution for the dry timber density

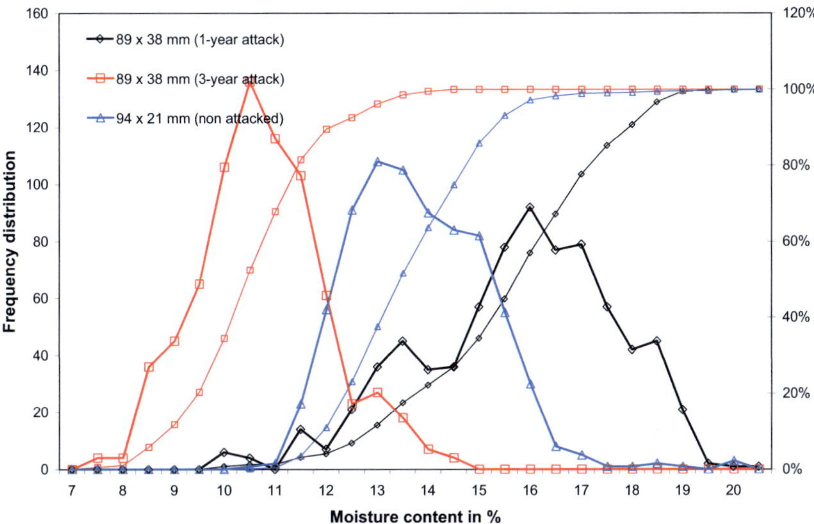

Fig. 10   Frequency distribution for the moisture content

In Fig. 11 the relationship between the dynamic modulus of elasticity and the dry timber density is given. Obviously, the MOE is not influenced by the strength of the MPB attack. The MOE for timber which was attacked by the MPB timber for one year is quite equal to the MOE for timber which was attacked by the MPB for three years. The MOE for the non attacked timber is smaller than for the timber attacked by the MPB. The low value of the MOE can be explained by the low grade of this timber. The frequency distribution of the MOE is displayed in Fig. 12. Considering all results, the main results are summarized in Table 1.

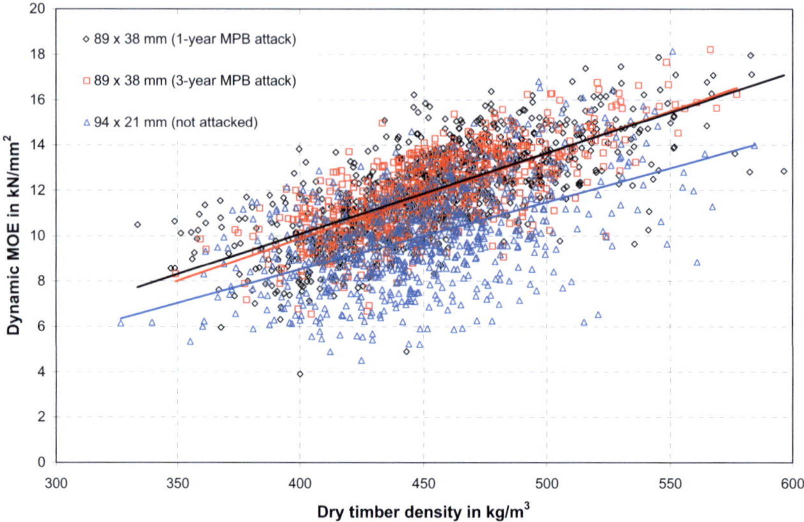

Fig. 11   Relation between the dynamic MOE and the dry timber density

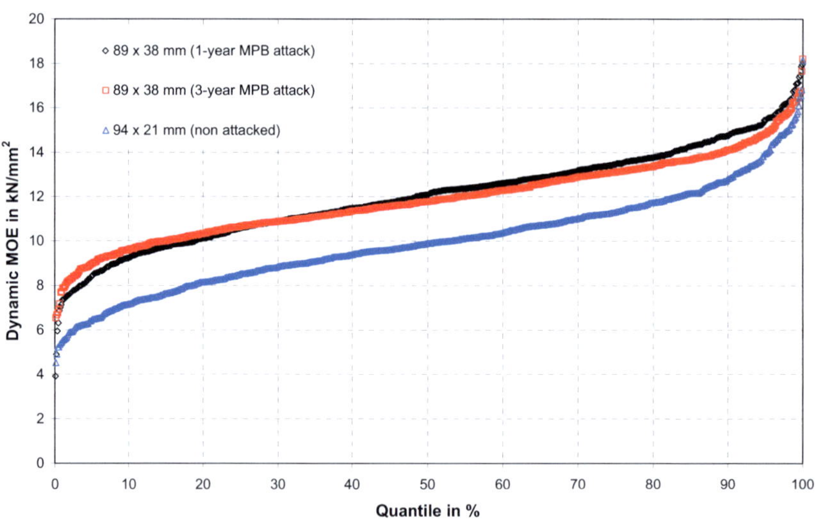

Fig. 12   Frequency distribution for the MOE

Table 1   Main parameters considering the pre-grading

| Series | Dimensions | Timber | n | $E_{5\%}$ | $E_{50\%}$ | $\rho_{5\%}$ | $u_{50\%}$ |
			[-]	[N/mm^2]	[N/mm^2]	[kg/m^3]	[%]
1	89 x 38 mm	1 year MPB attack	756	8,39	12,1	387	15,7
		3 year MPB attack	756	8,97	11,8	402	10,9
2	94 x 21 mm	non attacked timber	756	6,48	9,89	388	13,9

Taking into account the German grading classes for soft wood, timber from series 1 can be classified into the German grading class C27/C30. Against it, timber from series 2 can be classified into the German grading class C20/C22. Both grading classes are usually used for European cross laminated timber panels.

The determination of the timber properties took place in the laboratories of the Department of Wood Science at the University of British Columbia. Afterwards, the timber was transported to the former "CANFOR" research laboratories in New Westminster, British Columbia, Canada, where the specimens were manufactured.

Because not all raw materials were used for the CLT and DLT beams, an additional end-grading at the laboratories in New-Westminster was done. The first results from the pre-grading demonstrate that the duration of the MPB attack does not affect the MOE and the density of the timber. Hence, the end-selection was reduced to only two grading types, considering attacked timber in general (species 1) and non-attacked timber (species 2).

The CLT and DLT beams were produced using 800 timber beams from species 1 and 694 timber beams from species 2. The properties of the finally used timber are summarized in the following figures. Fig. 13 shows the frequency distribution for the dry timber density for beams with 89 x 38 mm (series 1) and for beams with 94 x 21 mm (series 2). The relation between the dynamic modulus of elasticity and the dry timber density for both timber series is displayed in Fig. 14. In Fig. 15 the frequency distribution for the modulus of elasticity is displayed. Comparing the results from the end-grading with those from the pre-grading no significant differences can be noticed. For this reason the main parameters from the end-grading (Table 2) are quite similar to those from the pre-grading (Table 1).

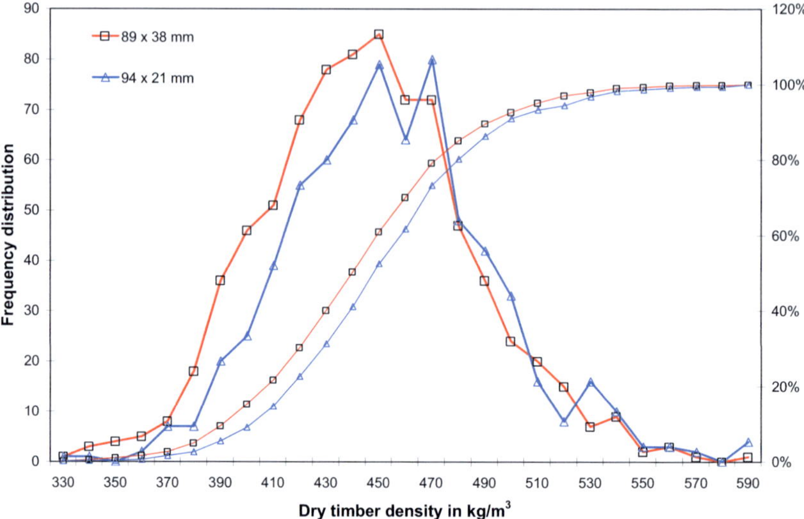

Fig. 13   Frequency distribution for the dry timber density

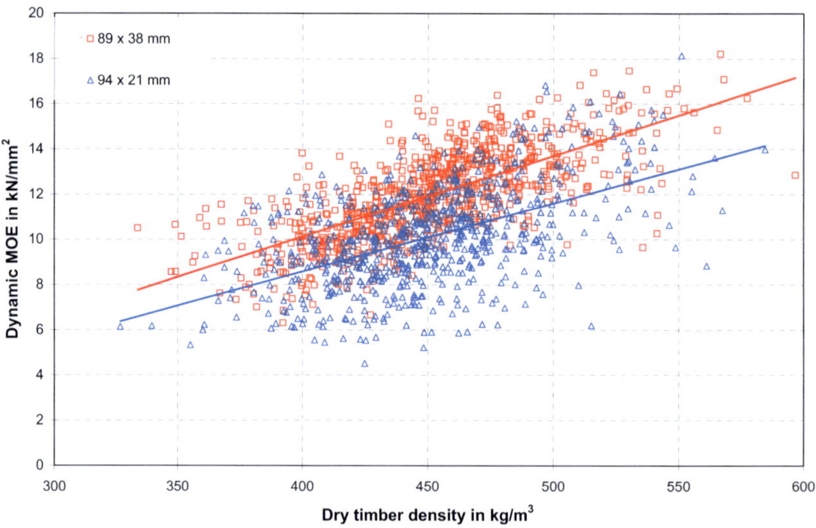

Fig. 14   Relation between the dynamic MOE and the dry timber density

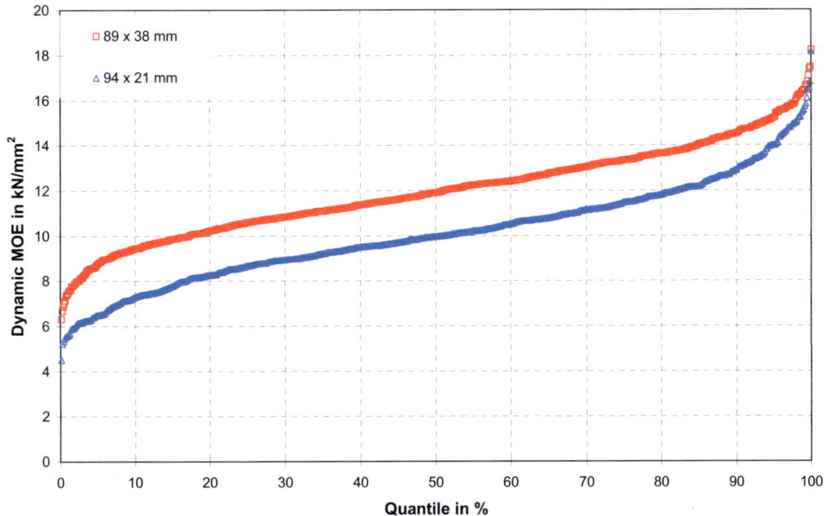

Fig. 15   Frequency distribution for the MOE

Table 2   Main parameters for the end-grading

Series	Dimensions	Timber	n [-]	$E_{5\%}$ [N/mm²]	$E_{50\%}$ [N/mm²]	$\rho_{5\%}$ [kg/m³]	$u_{50\%}$ [%]
1	89 x 38 mm	1 year MPB attack	518	8,79	11,9	390	13,9
		3 year MPB attack	282				
2	94 x 21 mm	non attacked timber	694	6,49	9,97	388	13,6

3.3   Production of CLT beams

Using the end-graded timber, 60 CLT and DLT beams were produced. As previously mentioned, the maximum press load is limited to 850 kN. Considering the maximum dimensions of the press (4 x 2 m²) and the maximum load, only a total pressure of 0.11 N/mm² can be applied on full elements. However, most of the glue manufacturers recommend a pressure of 1 N/mm² for their glue to assure an adequate bonding between the timber members. Therefore, it was not possible to produce whole panels with maximum dimensions and afterwards to cut them to the necessary geometry of the beams. Unfortunately, only two beams were produced in one run, which was pretty inefficient. In practice, the beams can be cut out of whole panels which are previously manufactured in one run.

Following 60 CLT and DLT beams were produced in 30 runs using the end-graded raw material:

Species 1.1: 12 short beams with L x H x B = 244 x 37 x 14 cm^3, 90°, 5-ply (Fig. 47)
Species 1.2: 12 long beams with   L x H x B = 427 x 27 x 14 cm^3, 90°, 5-ply (Fig. 48)
Species 2.1: 8 short beams with   L x H x B = 244 x 37 x 14 cm^3, 45°, 5-ply (Fig. 49)
Species 2.2: 10 short beams with L x H x B = 244 x 37 x 11 cm^3, 45°, 4-ply (Fig. 50)
Species 2.3: 8 long beams with   L x H x B = 427 x 27 x 14 cm^3, 45°, 5-ply (Fig. 51)
Species 2.4: 10 long beams with   L x H x B = 427 x 27 x 11 cm^3, 45°, 4-ply (Fig. 52)

To enforce shear failure, short beams (species 1.1, 2.1, and 2.2) were produced. To enforce bending failure, long beams (species 1.2, 2.3 and 2.4) were produced. Species 1.1 and 1.2 were made using crosswise orientated boards. Thereby, three layers were orientated in the beam axis direction while two layers were orientated perpendicular to the beam axis direction. Species 2.1, 2.2, 2.3 and 2.4 were made using diagonal orientated layers. Either specimens with five layers (2.1 and 2.3) or specimens with four layers (2.2 and 2.4) were manufactured. Those DLT beams with five layers were made using three parallel and two diagonal layers. Thereby, in between the two diagonal layers, which were orientated in the opposite direction to each other, a parallel layer was placed. DLT beams with four layers were made with two outer parallel and two inner diagonal layers. Thereby, the diagonal layers were orientated in the opposite direction and side by side.

For all specimens, timber from series 1 (89 x 38 mm^2) was used for the parallel layers. All perpendicular and diagonal orientated layers were made using timber from series 2 (94 x 21 mm^2). Due to the limited length of 8' (2.44 m) of the raw material, the boards for the parallel layers in longer beams were connected together using finger joints to increase the length to 14' (4,27 m). Therefore, totally 480 boards from series 1 were finger jointed together to 240 boards with a total length of 14' (4.27 m). The location of the finger joints in long CLT and DLT beams is marked in Fig. 48, Fig. 51 and Fig. 52 using a red line. Within the next manufacturing step, all timber beams were planed over their wide surfaces. The narrow surfaces weren't planed. Timber from series 1 was planed to a new thickness of 35 mm while timber from series 2 was planed to a new thickness of 17.5 mm. Due to the reduced pressure of the press simply beams instead of panels were produced. Therefore, it was necessary to chop all the beams from series 2 used for the perpendicular and diagonal layers to their final length. The geometry and number of pieces which is necessary for any beam can be taken from Fig. 47 to Fig. 52. All beams were glued using a Phenol-Resorcinol-Formaldehyde-glue provided by HEXION. Thereby, CASCOPHEN AG-5635Q was mixed together with CASCOSET FM-6310L in a proportion of 2.33. The processing time for the glue is given by the manufacturer to 25 to 30 minutes. Two beams remained for about 45 – 90 minutes at the same time in the heated press. The oil temperature was 90°C. Considering the maximum hydraulic jack load of 850 kN and the geometry of the beams, long beams were pressed with 0.4 N/mm^2, while short beams were pressed with 0.5 N/mm^2. Although this pressure is definitely lower than the recommended pressure by the glue manufacturer, almost no glue failure was noticed in the tests.

The manufacturing process is documented in Fig. 53 to Fig. 66. The assembling time for two beams was limited to 25 to 30 minutes due to the processing time for the glue. To accelerate the production process, a self-made frame was used for the beams. This frame was used as well to hold the beam members together during the pressing and hardening process. The beams were assembled together layer by layer. Between each layer glue was applied only on one surface using a commercial painter roller.

During the pre-grading all single timber members were numbered. Those numbers consists information about the dynamic modulus of elasticity and the dry timber density for each timber member. The modulus of elasticity and the dry timber density for the single timber members are listed in Table 14 to Table 19. To locate the single timber members in the corresponding beam use Fig. 47 to Fig. 52.

3.4   Testing and test results

The final cutting and testing of the manufactured 60 beams took place at the University of British Columbia. First, 30 beams were tested in bending without any changing. The remaining 30 beams were used to demonstrate the effect of the diagonal and perpendicular layers. Therefore, rectangular holes and end notches were cut into the remaining 30 beams. Those singularities affect the load-carrying capacity of beams due to tensile stresses perpendicular to the grain direction. Geometrical identical solid wood and glulam beams would split at lower load-carrying capacities. Against it, for CLT and DLT beams with holes and notches higher load-carrying capacities were expected.

3.4.1   Short beams without singularities

15 short CLT and DLT beams were tested in bending according to the test set-up in Fig. 16 to determine the bending stiffness and the load-carrying capacity. Thereby two single loads were applied on the single supported beam. Thereby, the load, the displacement of the hydraulic jacks, the total and the local vertical displacements were measured. The specimens were loaded until failure. The testing speed was determined to get the first failure in between 5 ± 1 minute. The total and local displacements were measured using displacement transducers. Those were removed at approximately 70% of the estimated load-carrying capacity and short before the beams failed to avoid transducer damage.

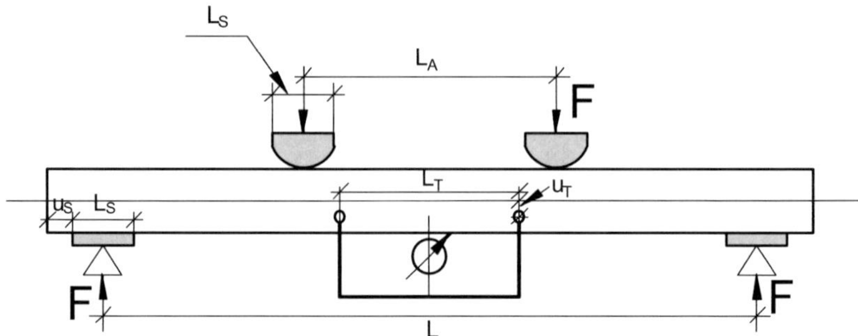

Fig. 16   Test set-up for short beams

Following dimensions were used for the test set-up:

L = 2100 mm, L_A = 800 mm, L_T = 600 mm, u_T = 10 mm, L_S = 300 mm, u_S = 20 mm. All tested short beams were H = 360 mm in the height. Beams with five layers from series 1-1 and 2-1 were B = 140 mm in the width while beams with four layers from series 2-2 were B = 105 mm in the width. As previously mentioned, parallel orientated boards were B_{par} = 35 mm thick while perpendicular or diagonal orientated boards were B_{perp} = 17.5 mm in the thickness.

Table 3 contents the main results for short beams without singularities. In the first row the specimen number is displayed, where the first two numbers represent the species according to Fig. 47 to Fig. 52. In the upper third of the table, test results for short CLT beams with five plies and with perpendicular to the beam axis orientated layers (species 1-1) are given. Results for short DLT beams with fife plies and with diagonal orientated layers (species 2-1) are given in the middle third of the table. Finally, in the lower third of the table the results for short DLT beams with four plies and with diagonal orientated layers (species 2-2) are displayed. The ultimate load F_{total} for both hydraulic jacks is given in the second row. In row three to five the bending stresses σ_m, shear stresses τ_v and the local modulus of elasticity E_0 are given. These values were calculated according to the beam theory at the ultimate load considering the total beam thickness, which includes the parallel and perpendicular layers. In row six to eight the corresponding net values are given. These values were calculated using the ultimate load and only the thickness of the parallel layers. Thereby, perpendicular to the beam axis orientated layers were not taken into account.

Table 3   Main results for short beams without singularities

specimen	F_{total}	σ_m	τ_v	E_0	$\sigma_{m,net}$	$\tau_{v,net}$	$E_{0,net}$
	[kN]	[N/mm^2]	[N/mm^2]	[N/mm^2]	[N/mm^2]	[N/mm^2]	[N/mm^2]
1-1-1	303	32,5	4,51	8191	43,4	6,01	10921
1-1-2	292	31,4	4,35	8427	41,9	5,80	11236
1-1-5	248	26,7	3,69	7246	35,5	4,92	9662
1-1-6	242	26,0	3,60	7872	34,6	4,80	10497
1-1-8	263	28,2	3,91	7176	37,7	5,22	9568
mean	**270**	**29,0**	**4,01**	**7782**	**38,6**	**5,35**	**10376**
2-1-1	291	31,2	4,33	8087	41,7	5,77	10782
2-1-2	287	30,8	4,27	6810	41,1	5,70	9080
2-1-4	320	34,3	4,76	8471	45,8	6,34	11295
2-1-5	305	32,8	4,54	7802	43,7	6,05	10402
2-1-7	301	32,3	4,48	7678	43,1	5,97	10237
mean	**301**	**32,3**	**4,47**	**7769**	**43,1**	**5,96**	**10359**
2-2-1	248	35,5	4,92	6008	53,3	7,38	9012
2-2-2	191	27,4	3,79	7500	41,1	5,69	11250
2-2-3	272	38,9	5,39	7897	58,4	8,08	11845
2-2-4	229	32,9	4,55	7785	49,3	6,82	11678
2-2-5	207	29,6	4,10	6659	44,4	6,15	9989
mean	**229**	**32,9**	**4,55**	**7170**	**49,3**	**6,83**	**10755**

The load-displacement behaviour for all short beams is displayed in Fig. 67 to Fig. 69. Fig. 67 shows the relationship between the total load considering both hydraulic jacks and the displacement of the hydraulic jacks. The relationship between the total load considering both hydraulic jacks and the total and local beam deflection is given in Fig. 68 and Fig. 69. The total deflection was determined considering the beam span. In contrast, the local deflection was determined between the two single loads on a length L$_T$. The modulus of elasticity was calculated according to equation (2) using the local deflection. Between the single loads, no shear occurs. Therefore, the local modulus of elasticity doesn't consider the shear component.

$$E_{0,(net)} = \frac{(L - L_A) \cdot L_A^2 \cdot F_{total}}{32 \cdot I_{(net)} \cdot \Delta u_T}$$

(2)

With

L$_A$ Length between single loads

Δu_T Relative displacement in local area

F_{total} Total load considering both hydraulic jacks

$I_{(net)}$ (Net) moment of inertia

Short beams were tested to enforce shear failure. In contrast to the expectation, some beams failed in bending (left in Fig. 70) while other beams failed in shear (Fig. 70 to Fig. 72). Bending failure is characterised by tensile failure of one of the bottom and in the beam axis orientated boards. Against it, shear failure is characterised by delaminating of one of the outer and in the beam axis orientated boards. The shear failure occurred at the beam ends while bending failure occurred between the single loads. Some beams failed due to a combined failure due to shear and bending. The reason for the combined failure was that the bending and shear stresses, which were calculated according to the beam theory, were very close to their strength properties (Table 3). Although the failure was different, a difference between the different beam types can be showed taking into account the results in Table 3.

First, the local net modulus of elasticity is nearly similar for CLT and DLT beams. Hence, the beam type does not affect the bending stiffness significantly. However, the test results show differences for the net stresses and hence for the load-carrying capacities. The lowest average bending and shear stresses were reached for CLT beams with five plies and with perpendicular to the beam axis orientated layers (species 1-1). For DLT beams with diagonal layers and five plies (species 2-1) the average stresses were about 12% higher. Finally, the best result was reached for DLT beams with diagonal layers and with four plies. Considering only the parallel layers and the net cross-section, the increase in strength compared to CLT beams was 28%. But even in case of taking into account the total cross-section including parallel and diagonal layers, the stresses for DLT beams with four plies were about 14% higher compared to the CLT beams. Thereby, for DLT beams with four plies the ratio between the thickness of the parallel layers and the total beam thickness is only 0.67. Against it, the same ratio for CLT beams is 0.75. Obviously, DLT beams with two diagonal layers, which are orientated in the opposite direction to each other and side by side are much better than CLT beams. Assuming approximately equal strength properties for the single timber members and taking into account the test results, the stresses for DLT beams should be smaller than those calculated according to the beam theory. This assumption will be investigated using the finite element analysis.

3.4.2   Long beams without singularities

Bending tests on 15 long CLT and DLT beams according to the test set-up in Fig. 17 were done to determine the bending stiffness and the load-carrying capacity. The test set-up used for long beams is similar to those used for short beams.

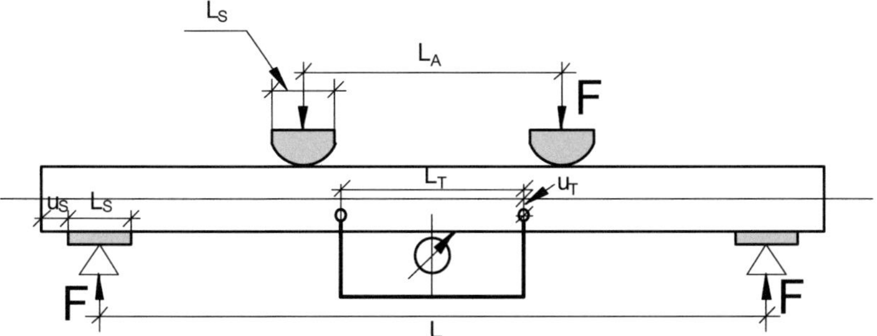

Fig. 17   Test set-up for long beams

For long beams following dimensions were used:

L = 3800 mm, L_A = 1300 mm, L_T = 1000 mm, u_T = 0 mm, L_S = 300 mm, u_S = 85 mm

All tested long beams were H = 270 mm in the height. The thickness for specimens from test series 1-2 and 2-3 with five plies was B = 140 mm while the thickness for specimens from test series 2-4 with four plies was B = 105 mm. As already mentioned, parallel to the beam axis orientated boards were B_{par} = 35 mm thick. The thickness of the boards used for the diagonal or perpendicular layers was B_{perp} = 17.5 mm.

Table 4 consider the main results for long beams without singularities. The layout in Table 4 is similar to those for short beams. In the first row the specimen number is displayed. The first numbers represent the species as displayed in Fig. 47 to Fig. 52. The upper third of the table deals with results for long CLT beams with five plies and with perpendicular to the beam axis orientated layers (species 1-2). Results for long DLT beams with five plies and with diagonal to the beam axis orientated layers (species 2-3) are given in the middle third of the table. Finally, in the lower third of the table, results for long DLT beams with four plies and with two side by side diagonal orientated layers (species 2-4) are displayed. The ultimate load F_{total} considering both hydraulic jacks is given in the second row. In row three to five the bending stresses σ_m, shear stresses τ_v and the local modulus of elasticity E_0 are given. These values were calculated at the ultimate load using the total beam thickness including the parallel and perpendicular layers. The bending and shear stresses were calculated using the beam theory. The local modulus of elasticity was calculated according to equation (2). In row six to eight, net values are given. These values were calculated using only the thickness of the parallel to the beams axis orientated layers.

Table 4   Main results for long beams without singularities

specimen	F_{total}	σ_m	τ_v	E_0	$\sigma_{m,net}$	$\tau_{v,net}$	$E_{0,net}$
	[kN]	[N/mm^2]	[N/mm^2]	[N/mm^2]	[N/mm^2]	[N/mm^2]	[N/mm^2]
1-2-8	92,1	33,8	1,83	9026	45,1	2,44	12034
1-2-9	105	38,6	2,08	10007	51,4	2,78	13343
1-2-10	93,8	34,5	1,86	10352	46,0	2,48	13803
1-2-11	80,2	29,5	1,59	7045	39,3	2,12	9393
1-2-12	87,8	32,3	1,74	9603	43,0	2,32	12804
mean	**91,8**	**33,7**	**1,82**	**9206**	**45,0**	**2,43**	**12275**
2-3-3	79,7	29,3	1,58	9408	39,0	2,11	12544
2-3-4	81,4	29,9	1,62	9128	39,9	2,15	12171
2-3-5	91,5	33,6	1,82	8400	44,8	2,42	11201
2-3-7	95,4	35,1	1,89	9217	46,7	2,52	12289
2-3-8	87,2	32,0	1,73	9221	42,7	2,31	12294
mean	**87,1**	**32,0**	**1,73**	**9075**	**42,6**	**2,30**	**12100**
2-4-2	68,6	33,6	1,82	9249	50,4	2,72	13874
2-4-4	52,6	25,8	1,39	8223	38,6	2,09	12334
2-4-8	56,4	27,6	1,49	8693	41,4	2,24	13039
2-4-9	57,4	28,1	1,52	8390	42,1	2,28	12585
2-4-10	56,5	27,7	1,49	8804	41,5	2,24	13206
mean	**58,3**	**28,6**	**1,54**	**8672**	**42,8**	**2,31**	**13008**

The corresponding load-displacement behaviours for all long beams are displayed in Fig. 73 to Fig. 75. The first diagram represents the relationship between the hydraulic jack displacement and the total load. The other both diagrams represent the relationship between the total load and the total or local deflection of the single supported beam. Like for short beams, to protect the displacement transducers, they were removed shortly before the failure occurred.

Long beams were tested to enforce bending failure. As expected, all long beams failed in bending (Fig. 76 to Fig. 78). Bending failure is characterised by tensile failure of at least one of the bottom and in the beam axis orientated boards.

Compared to the test results for short beams, the local modulus of elasticity for all beams was much higher. However, like for the short beams, no significant increase in bending stiffness was reached using DLT beams instead of CLT beams. Obviously, the beam type does not affect the bending stiffness significantly. In contrast to the test results for short beams, no significant difference in load-carrying capacity and hence in bending strength was observed. It seems that the beam type does not affect the load-displacement behaviour for long beams. Strictly speaking, the average net stresses for DLT beams were even 5% smaller than the average net stresses for CLT beams. For better qualification numerical investigations using the finite element analysis are necessary.

3.4.3   Short beams with notches

Due to the reinforcing layers, CLT and DLT are more efficient in beams with singularities than in common beams loaded in bending. Those singularities can be notched ends and round or rectangular holes. For this reason, tests on beams with notched ends and holes were done. It was expected, that the performance of CLT and DLT beams with notched ends and holes is much better than those of geometrical identical solid wood or glulam beams with similar singularities.

Short beams were used to demonstrate the effect for notches while long beams were used for beams with holes. At first, 15 short CLT and DLT beams were prepared using the HUNDEGGER K2 CNC machine and afterwards, they were tested in bending to determine the bending stiffness and the load-carrying capacity. Using the HUNDEGGER K2 CNC machine, notches for short beams were cut out. The beams were tested in bending according to the test set-up in Fig. 18.

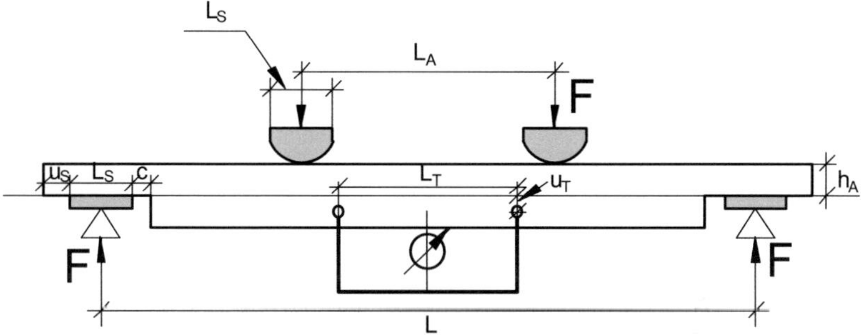

Fig. 18   Test set-up for short beams with notches

The dimensions for short beams with notches were similar to those for short beams without singularities:

$L = 2100$ mm, $L_A = 800$ mm, $L_T = 600$ mm, $u_T = -10$ mm, $L_S = 300$ mm, $u_S = 0$ mm

The short and end notched beams were $H = 360$ mm in the height. Beams from test series 1-1 and 2-1 with five plies were $B = 140$ mm in the width while beams from test series 2-2 were $B = 105$ mm in the width. Thereby, parallel to the beam axis orientated boards were $B_{par} = 35$ mm thick. The thickness of the perpendicular or diagonal to the beam axis orientated layers was $B_{perp} = 17.5$ mm.

The spacing between the edge of the beam support and the edge of the notch was chosen to $c = 50$ mm. The remaining beam height at the beam support was chosen to $h_A = 0.5 \cdot H = 180$ mm. Hence, the ratio between the remaining beam height h_A and the total beam height H was $\alpha = 0.5$. This is the maximum permitted ratio according to [1]. With increasing ratio α the load-carrying capacity degreases. Therefore, for $\alpha = 0.5$ the lowest load-carrying capacity is expected.

According to [1] the shear stress at a notched beam support can be calculated using equation (3) to (5). Equation (3) calculates the shear stress around the notched beam support taking into account the remaining beam height h_A and a reduction factor k_v. As long as the reduction factor k_v equals 1, no splitting and hence no failure around the notched edge occurs. It is expected, that for notched beam supports made using CLT or DLT those kind of failure does not occur.

$$\tau_v = 1,5 \cdot \frac{V}{B \cdot h_A} \cdot \frac{1}{k_v} \tag{3}$$

With

$$k_v = \min \left\{ \begin{array}{c} 1 \\ k_{90} \cdot k_\varepsilon \end{array} \right\} \quad \text{with} \quad k_\varepsilon = 1 \quad \text{for} \quad \varepsilon = 90° \tag{4}$$

And with

$$k_{90} = \frac{k_n}{\sqrt{H} \cdot \left(\sqrt{\alpha \cdot (1-\alpha)} + 0,8 \cdot \frac{c}{H} \cdot \sqrt{\frac{1}{\alpha} - \alpha^2} \right)} \tag{5}$$

For beams with an equal geometry but made in glulam instead of CLT or DLT, the factor k_v can be calculated to $k_v = 0.53$ (using $\alpha = 0.5$, $c = 50$ mm, $H = 360$ mm and $k_n = 6.5$ for glulam). For this reason, splitting around the notched edge should occur at a calculated shear stress which is about half of the shear strength for timber. However, the calculated shear stresses for the investigated CLT and DLT beams were much higher than the half of the shear strength and splitting likewise observed for common glulam beams didn't occur. Table 5 contains the mentioned shear stresses in row seven. These shear stresses were calculated considering the remaining beam height h_A which was assumed as 50% of the total beam height. The average shear stress for CLT beams is $\tau_{v,net} = 6.14$ N/mm^2. For DLT beams with five plies the average shear stress is $\tau_{v,net} = 5.45$ N/mm^2, while for the DLT beams with four plies the average shear stress was calculated to $\tau_{v,net} = 6.87$ N/mm^2. Those values are quite similar to the shear strength for the used timber. As expected, for the investigated CLT and DLT beams with notched ends no splitting failure occurred. These results show that the performance of CLT and DLT beams with notched ends is much better that those for similar beams made using solid wood or glulam.

Table 5 contains more results. In row three to five the bending stresses σ_m, shear stresses τ_v and the local modulus of elasticity E_0 are given. These values were determined using the beam theory considering the total beam thickness including the parallel, perpendicular and diagonal to the beam axis orientated layers. Row six to eight contains values for the bending stresses, the shear stresses and the local modulus of elasticity. These values were calculated considering only the parallel to the beam axis orientated layers. Both values, the local modulus of elasticity in row six and eight, were calculated using equation (2). Likewise for short and long beams without singularities, the bending stresses and the modulus of elasticity were calculated taking into account the total height of the beams. Against it, the shear stresses were calculated considering the remaining beam height h_A which was assumed as 50% of the total beam height. Thereby the shear stresses in Table 5 were not reduced as ruled in [1] using the factor k_v. It was assumed, that splitting at the notched ends doesn't occur due to sufficient reinforcing by the perpendicular or diagonal layers. The full diagrams considering the load-displacement behaviour are displayed in Fig. 79 to Fig. 81.

Table 5   Main results for short beams with notches

specimen	F_{total}	σ_m	τ_v	E_0	$\sigma_{m,net}$	$\tau_{v,net}$	$E_{0,net}$
	[kN]	[N/mm^2]	[N/mm^2]	[N/mm^2]	[N/mm^2]	[N/mm^2]	[N/mm^2]
1-1-3	153	16,5	4,56	8779	22,0	6,08	11705
1-1-4	178	19,1	5,30	7940	25,5	7,07	10587
1-1-7	124	13,3	3,69	7361	17,8	4,93	9815
1-1-9	125	13,5	3,73	8956	18,0	4,97	11942
1-1-10	163	17,5	4,85	7361	23,3	6,46	9815
1-1-11	182	19,5	5,41	8888	26,0	7,21	11851
1-1-12	158	17,0	4,71	7929	22,7	6,28	10573
mean	**155**	**16,6**	**4,61**	**8174**	**22,2**	**6,14**	**10898**
2-1-3	104	11,2	3,09	9711	14,9	4,13	12948
2-1-6	134	14,4	3,99	8811	19,2	5,32	11748
2-1-8	174	18,7	5,18	8979	25,0	6,91	11971
mean	**137**	**14,8**	**4,09**	**9167**	**19,7**	**5,45**	**12223**
2-2-6	113	16,2	4,48	7365	24,3	6,72	11048
2-2-7	130	18,7	5,18	8432	28,1	7,77	12648
2-2-8	109	15,7	4,34	8097	23,5	6,51	12146
2-2-9	-	-	-	-	-	-	-
2-2-10	109	15,6	4,31	7801	23,4	6,47	11702
mean	**115**	**16,5**	**4,58**	**7924**	**24,8**	**6,87**	**11886**

Although the results in Table 5 are difficult to understand, the observed failures can help to qualify the different beam types. Two failures were observed for this test series. Thereby, some beams failed in pure shear which occurred at the beam end above the notched area. This failure was characterised by delaminating of at least one of the outer boards, which are orientated in the beam axis direction. Thereby, no failure in the notched edge was observed. For this reason, the calculated shear stresses in row seven in Table 5 according to the beam theory are nearly similar to the shear strength of the single timber members. Against it, for some beams the failure occurred in the area of the notched edges. Thereby, either the reinforcing inner layers failed in tensile or pure glue failure between the parallel orientated and the reinforcing layers occurred. This kind of failure is documented in Fig. 82 to Fig. 84.

This second failure occurs, when the tensile load perpendicular to the grain direction at the notched edge is smaller than the load-carrying capacity of the reinforcing layer. This tensile load perpendicular to the grain direction can be calculated according to [1] using following equation:

$$F_{t,90} = 1,3 \cdot V \cdot \left[3 \cdot (1-\alpha)^2 - 2 \cdot (1-\alpha)^3 \right] \qquad (6)$$

Thereby the tensile load depends on the ratio α and on the shear load. Considering the average total loads for CLT and DLT beams, the tensile load can be calculated to $F_{t,90}$ = 50.4 kN for CLT beams, to $F_{t,90}$ = 44.5 kN for DLT beams with five plies and to $F_{t,90}$ = 37.4 kN for DLT beams with four plies. Taking into account these results, perpendicular to the beam axis orientated reinforcing layers provide obviously the best reinforcing method. Tensile loads in the notched area occur only locally and close to the notched end. For this reason only the first board, which is located close to the notched edge, is loaded by $F_{t,90}$. Considering two perpendicular layers and a cross-section of 94 x 17.5 mm^2 for each perpendicular to the beam axis orientated board, the first boards in each layer were loaded by a tensile stress which is $f_{t,0}$ = 15.3 N/mm^2. Tensile failure occurred, because the maximum tensile stress in grain direction $f_{t,0}$ = 15.3 N/mm^2 was nearly similar to the characteristic tensile strength in grain direction. As previously mentioned, the perpendicular to the beam axis orientated boards can be graded into the class C20/C22 according to [1]. The tensile strength in grain direction for C20/C22 is $f_{t,0,k}$ = 12 – 13 N/mm^2.

As long as the load-carrying capacity of a beam with notched ends is governed by the load-carrying capacity of the notched beam supports, CLT should be preferred to DLT. The reinforcing layers in CLT are orientated perpendicular to the beam axis direction and hence in the direction of the tensile stresses, which occur at the notched beam support. Against it, the reinforcing layers in DLT are orientated in diagonal direction and hence they aren't orientated in the direction of the tensile stresses. The load transfer between the diagonal layers and the tensile load perpendicular to the beam axis direction can be regarded as a truss system. Therefore, with degreasing inclination of the diagonal layers, the reinforcing effect degreases.

3.4.4 Long beams with holes

Finally, 15 long CLT and DLT beams with holes were first cut using the HUNDEGGER K2 machine and afterwards they were tested in bending to determine the bending stiffness and the load-carrying capacity. The holes geometry was determined according to [1]. Thereby the height of a hole is limited to 40% of the total height of the beam while the length of the hole is limited to the total beam height. For the specimens almost the maximum holes dimensions were chosen because with increasing dimensions the load-carrying capacity degreases. The beams were tested in bending according to the test set-up in Fig. 19.

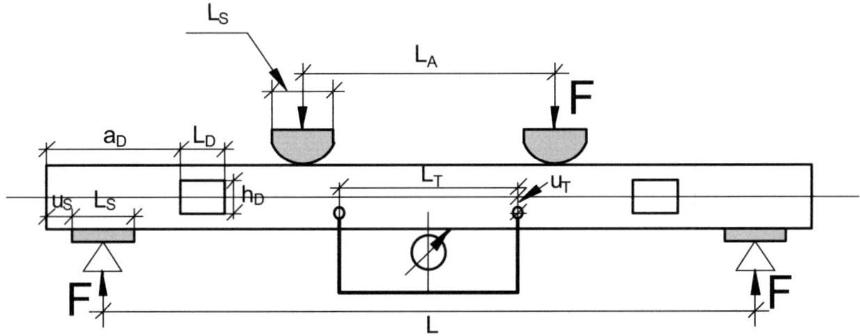

Fig. 19   Test set-up for long beams with holes

The outer dimensions for long beams with rectangular holes were similar to those for long beams without singularities:

L = 3800 mm, L_A = 1300 mm, L_T = 1000 mm, u_T = 0 mm, L_S = 300 mm, u_S = 50 mm

The tested long beams with holes were H = 270 mm in the height. Beams from test series 1-2 and 2-3 were B = 140 mm in the width. Against it, beams from test series 2-4 were B = 105 mm in the width. The thickness of the parallel orientated boards was B_{par} = 35 mm while the perpendicular or diagonal to the beam axis orientated boards were B_{perp} = 17.5 mm thick.

The hole edge distance to the beam end was a_D = 690 mm. The holes were placed in the middle between the beam support and the neighbouring single load to avoid an influence of the stresses which occur around the loading area. The length of the holes was chosen to L_D = H = 270 mm. The holes were h_D = 100 mm in the height, which is about 37% of the beam height. The beams were designed in that way, that similar solid wood or glulam beams should fail due to tensile stresses perpendicular to the grain direction around the holes. Thereby, the loads should be significantly smaller than those for similar beams without holes. However, it was expected, that CLT and DLT beams don't behave like solid wood or glulam. Hence, no splitting failure around the holes for CLT and DLT beams was expected.

In Table 6 the main results for long beams with holes are given. In row two the total load for both hydraulic jacks is given. The corresponding bending and shear stresses as well as the modulus of elasticity were calculated using the beam theory. Row three to five contains the values which consider the total beam thickness. In row six to eight values are given, which consider only the parallel to the beam axis orientated layers. The maximum bending stresses occur in between the single loads while maximum shear stresses occur at the holes. Hence, bending stresses were calculated considering the total beam height while for the maximum shear stresses only the net height at the beam holes was considered. Diagrams with the load-displacement behaviour for long beams with holes loaded in bending are displayed in Fig. 85 to Fig. 87.

Table 6   Main results for long beams with holes

specimen	F_{total}	σ_m	τ_v	E_0	$\sigma_{m,net}$	$\tau_{v,net}$	$E_{0,net}$
	[kN]	[N/mm^2]	[N/mm^2]	[N/mm^2]	[N/mm^2]	[N/mm^2]	[N/mm^2]
1-2-1	87,8	32,3	2,77	8642	43,0	3,69	11523
1-2-2	74,5	27,4	2,35	8851	36,5	3,13	11802
1-2-3	92,5	34,0	2,92	9514	45,3	3,89	12686
1-2-4	98,9	36,4	3,12	10629	48,5	4,16	14171
1-2-5	83,3	30,6	2,63	8674	40,8	3,50	11565
1-2-6	82,7	30,4	2,61	9588	40,5	3,48	12784
1-2-7	65,7	24,1	2,07	8223	32,2	2,76	10964
mean	**83,7**	**30,7**	**2,64**	**9160**	**41,0**	**3,51**	**12214**
2-3-1	78,3	28,8	2,47	8636	38,4	3,29	11514
2-3-2	94,2	34,6	2,97	9610	46,1	3,96	12814
2-3-6	85,7	31,5	2,70	8856	42,0	3,60	11808
mean	**86,0**	**31,6**	**2,71**	**9034**	**42,2**	**3,62**	**12045**
2-4-1	52,6	25,8	2,21	7592	38,7	3,32	11388
2-4-3	58,0	28,4	2,44	7902	42,6	3,65	11854
2-4-5	50,9	24,9	2,14	8313	37,4	3,21	12469
2-4-6	62,3	30,5	2,62	9321	45,8	3,92	13982
2-4-7	41,3	20,2	1,73	8199	30,3	2,60	12299
mean	**53,0**	**26,0**	**2,23**	**8266**	**39,0**	**3,34**	**12399**

Usually, the load-carrying capacity for beams with holes is governed by the load-carrying capacity around the hole. Thereby, high tensile and compression stresses perpendicular to the beam axis direction and high shear stresses occur around the holes. Similar beams with holes however made in solid wood or glulam would fail at lower loads. Against it, the load-carrying capacities for the investigated CLT and DLT beams were similar to those for CLT and DLT beams without holes and hence higher than the assumed load-carrying capacities for solid wood or glulam beams with holes.

The investigated CLT and DLT beams failed in bending. Neither tensile nor shear failure was observed around the holes. Thereby, the perpendicular or diagonal orientated reinforcing layers were able to prevent splitting and hence a brittle failure at lower load-carrying capacities. The typical observed failure in bending is displayed in Fig. 88 and Fig. 89.

According to [1] the load component perpendicular to the beam axis direction next to the holes edges can be calculated (see (7)).

$$F_{t,90} = \frac{V \cdot h_D}{4 \cdot H} \cdot \left[3 - \frac{h_D^2}{H^2} \right] + 0,008 \cdot \frac{M}{h_r} \qquad (7)$$

With

$$h_r = \frac{\left(H - h_D \right)}{2} \qquad \text{for rectangular holes located in the middle of the cross-section} \qquad (8)$$

The tensile load occurs locally. With increasing distance to the holes the tensile loads degreases. For this reason, only the first reinforcing board, which is orientated close to the holes, is mainly loaded by $F_{t,90}$. Considering the average load-carrying capacities for the tested long beams, the first board in CLT beams was loaded by $F_{t,90}$ = 13.5 kN. Against it, the next to the hole orientated board in DLT beams with five plies was loaded by $F_{t,90}$ = 13.9 kN and for DLT beams with four plies, the corresponding board was loaded by $F_{t,90}$ = 8.55 kN.

Taking into account the load component for CLT beams with $F_{t,90}$ = 13.5 kN and the cross-section of the reinforcing boards (94 x 17.5 mm^2), the first boards around the holes were loaded in tension by $f_{t,0}$ = 4.10 N/mm^2. The timber used for the reinforcing layers was graded into the class C20/C22 according to [1] with a characteristic tensile strength in grain direction of $f_{t,0,k}$ = 12 – 13 N/mm^2. Due to the low tensile stresses perpendicular to the beam axis, no failure around the holes was observed.

Although not observed, DLT could be better for beams with holes than CLT. The areas around the holes are loaded by tensile stresses perpendicular to the beam axis direction and by shear stresses. The resulting force which considers the shear and tensile force, is orientated diagonal to the beam axis direction and hence in the direction of the reinforcing layers in DLT elements. Similar results were made by Blass and Bejtka [13]. Glulam beams with holes were investigated which usually are prone to splitting due to tensile stresses perpendicular to the grain direction or due to shear stresses around the holes. Two reinforcing methods using self-tapping screws were investigated. One, where the screws were orientated perpendicular to the beam axis direction and one, where the screws were orientated at 45° to the beam axis direction. The second reinforcing method seemed to be better, because the screws were able to transfer both, high tensile loads perpendicular to the grain direction and high shear loads. Unfortunately, this effect couldn't be proved within this research project, because no one beam failed duel to shear stresses or high tensile stresses perpendicular to the beam axis direction and close to the holes area. Fortunately, it was demonstrated that beams with holes made using CLT or DLT usually are not prone to splitting around the holes area.

Likewise for long beams without holes, the investigated orientation of the reinforcing layers (45° or 90°) does not affect the bending strength and the MOE. Considering the test results, those values are nearly similar for DLT and CLT beams.

3.4.5   Assessment of the beam type

The aim of the tests was to demonstrate the reinforcing effect of CLT and DLT beams which are loaded in bending and by tensile stresses perpendicular to the grain direction. Furthermore, the tests were done to assess the type of the beam (CLT vs. DLT) in respect to the load-carrying behaviour. The results show that timber splitting, which usually occurs in common glulam or solid wood beams loaded in bending and by high tensile stresses perpendicular to the grain, can be prevented using CLT or DLT. Thereby, the tensile stresses perpendicular to the beam axis direction can be transferred by the reinforcing layers. However, the relationship between the beam type (CLT vs. DLT) and the bending stiffness could not be investigated so far. To determine the relationship between the different beams type (CLT vs. DLT) and the bending stiffness, the stiffness properties for all single timber members must be taken into account. In addition, all test results, those for beams without singularities and those for beams with singularities should be taken into account.

Subsequent, all results are taken into two groups to assess the different beam types in terms of the MOE. One group consist the local MOE for short beams and the other group consists the MOE for long beams. Thereby, results for beams without singularities and for beams with singularities are taken into one group. This can be done, because the investigated singularities like notches and holes, does not affect the local MOE. The local MOE was determined using the beam deflection in between the both single loads according to equation (2). In the following tables the local MOE values for all tested short beams (Table 7) and for all tested long beams (Table 8) are displayed. The MOE values in row two consider the total beam thickness. Against it, the MOE values in row three consider only the thickness of the parallel to the beam axis orientated layers. In the bottom lines mean values, standard deviations and 5%-quantile are given.

Table 7   Mean value, standard deviation and 5%-quantile for E_0 and $E_{0,net}$ for short beams

Short CLT beam - 5-ply				Short DLT beam - 5-ply				Short DLT beam - 4-ply		
specimen	E_0 [N/mm^2]	$E_{0,net}$ [N/mm^2]		specimen	E_0 [N/mm^2]	$E_{0,net}$ [N/mm^2]		specimen	E_0 [N/mm^2]	$E_{0,net}$ [N/mm^2]
1-1-1	8191	10921		2-1-1	8087	10782		2-2-1	6008	9012
1-1-2	8427	11236		2-1-2	6810	9080		2-2-2	7500	11250
1-1-3	8779	11705		2-1-3	9711	12948		2-2-3	7897	11845
1-1-4	7940	10587		2-1-4	8471	11295		2-2-4	7785	11678
1-1-5	7246	9662		2-1-5	7802	10402		2-2-5	6659	9989
1-1-6	7872	10497		2-1-6	8811	11748		2-2-6	7365	11048
1-1-7	7361	9815		2-1-7	7678	10237		2-2-7	8432	12648
1-1-8	7176	9568		2-1-8	8979	11971		2-2-8	8097	12146
1-1-9	8956	11942						2-2-9	-	-
1-1-10	7361	9815						2-2-10	7801	11702
1-1-11	8888	11851								
1-1-12	7929	10573								
mean value	8011	10681		mean value	8293	11058		mean value	7505	11258
st.deviation	648	864		st.deviation	898	1197		st.deviation	750	1126
5%-quantile	6850	9133		5%-quantile	6498	8664		5%-quantile	6034	9051

Table 8   Mean value, standard deviation and 5%-quantile for E_0 and $E_{0,net}$ for long beams

Long CLT beam - 5-ply				Long DLT beam - 5-ply				Long DLT beam - 4-ply		
specimen	E_0 [N/mm^2]	$E_{0,net}$ [N/mm^2]		specimen	E_0 [N/mm^2]	$E_{0,net}$ [N/mm^2]		specimen	E_0 [N/mm^2]	$E_{0,net}$ [N/mm^2]
1-2-1	8642	11523		2-3-1	8636	11514		2-4-1	7592	11388
1-2-2	8851	11802		2-3-2	9610	12814		2-4-2	9249	13874
1-2-3	9514	12686		2-3-3	9408	12544		2-4-3	7902	11854
1-2-4	10629	14171		2-3-4	9128	12171		2-4-4	8223	12334
1-2-5	8674	11565		2-3-5	8400	11201		2-4-5	8313	12469
1-2-6	9588	12784		2-3-6	8856	11808		2-4-6	9321	13982
1-2-7	8223	10964		2-3-7	9217	12289		2-4-7	8199	12299
1-2-8	9026	12034		2-3-8	9221	12294		2-4-8	8693	13039
1-2-9	10007	13343						2-4-9	8390	12585
1-2-10	10352	13803						2-4-10	8804	13206
1-2-11	7045	9393								
1-2-12	9603	12804								
mean value	9179	12239		mean value	9060	12080		mean value	8469	12703
st.deviation	989	1319		st.deviation	403	537		st.deviation	553	829
5%-quantile	7406	9875		5%-quantile	8254	11005		5%-quantile	7407	11111

MOE values ($E_{0,net}$) which consider only the thickness of parallel orientated layers can be used to compare the results for beams with five and with four plies. Taking into account the 5%-quantile values, the orientation of the reinforcing layers (90° or 45°) and hence the beam type (CLT vs. DLT) does not affect the local modulus of elasticity. The values for $E_{0,net,5\%}$ for CLT beams, DLT beams with five and four plies are nearly similar. However, a small difference in $E_{0,net,5\%}$ can be only noticed for long beams. The local MOE for beams with diagonal orientated reinforcing layers is about 12% higher than the corresponding local MOE for beams with perpendicular to the beam axis orientated reinforcing layers. Unfortunately, this comparison is quite meaningless, because the properties of the single timber members were not taken into account.

In the next step the previously determined MOE values for the single timber members, which are orientated in the beam axis direction, are taken into account to determine the relationship between the beam type and the local modulus of elasticity. For each tested beam the local modulus of elasticity was calculated considering the MOE values for the single timber members using equation (9). In this calculation, the diagonal or perpendicular to the beam axis orientated layers were neglected.

$$E_{0,cal} = \frac{\sum_{i=1}^{n}\left(E_i \cdot I_i + E_i \cdot A_i \cdot a_i^2\right)}{I_{net}} \tag{9}$$

With

E_i MOE value for the parallel orientated board i

I_i Moment of inertia for the parallel orientated board i

A_i Cross-section for the parallel orientated board i

a_i Distance between the centre of gravity for board i and the centre of gravity for the beam

I_{net} Moment of inertia for the beam considering only the parallel layers which act as a full plane

Using the beam configuration (Fig. 47 to Fig. 52) and the MOE values for each single and parallel to the beam axis orientated timber member (Table 14 to Table 19), the local modulus of elasticity $E_{0,cal}$ for the corresponding beam were calculated. Those values were compared with the local MOE values $E_{0,net}$ in Table 7 and Table 8.

In Fig. 90 to Fig. 95 the comparison between the calculated values ($E_{0,cal}$) and the test results ($E_{0,net}$) for each beam is given. Both values consider only the total thickness of the parallel to the beam axis orientated layers. The ratio between the test result $E_{0,net}$ and the calculated value $E_{0,cal}$ is displayed beneath each column pair. As long as this value is smaller than 100%, the parallel to the beam axis orientated layers can't be considered theoretically as a full plane.

The summarisation of all results is given in Fig. 20. Thereby, for each beam type the lowest, the highest and the average value for the ratio between the test result $E_{0,net}$ and the calculated value $E_{0,cal}$ is given. For short beams the average ratios are smaller than 100%. Against it, the average values for long beams are higher than 100%. Furthermore, for short and for long beams, the ratio between $E_{0,net}$ and $E_{0,cal}$ increases from CLT beams to DLT beams with five plies and finally to DLT beams with four plies. Obviously, the orientation of the reinforcing layers does affect the bending stiffness.

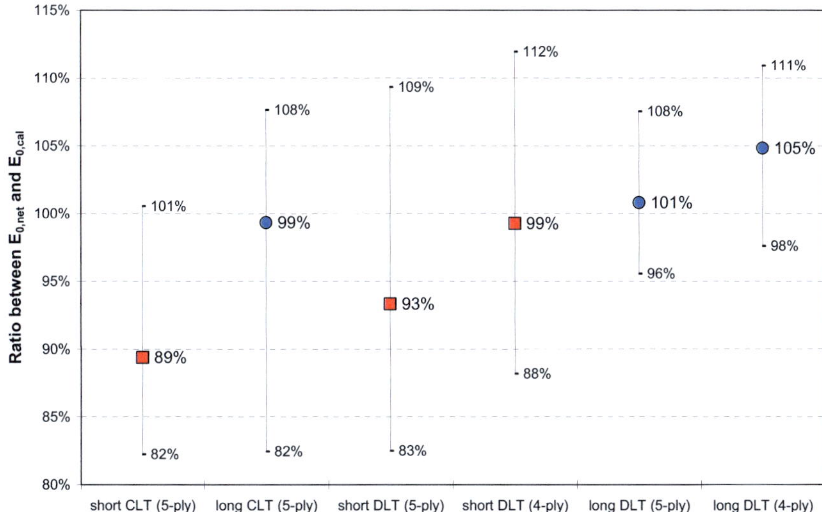

Fig. 20 $E_{0,net}$ versus $E_{0,cal}$ for each beam type

The previous investigations can be summarised as follows:
1. CLT and DLT beams can be used for beams which are loaded in bending and by tensile stresses perpendicular to the beam axis direction. Glulam and solid wood beams loaded by tensile stresses perpendicular to the beam axis direction or by high shear stresses are prone to splitting and hence they are less efficient compared to CLT and DLT beams.
2. In terms of shear stresses, DLT beams with diagonal to the beam axis orientated layers are better than CLT beams with perpendicular to the beam axis orientated layers. Hence, for short beams, where the load-carrying capacity is governed by shear, DLT instead of CLT for beams should be used.
3. Considering all results and the MOE values for the single timber members, local MOE values $E_{0,cal}$ for all the tested beams were calculated. These MOE values were compared to the test results (Fig. 20). Thereby a relationship between the beam type (CLT vs. DLT) and the local MOE and hence the bending stiffness was noticed. Obviously, the local MOE and hence the bending stiffness is higher for beams with diagonal orientated layers compared to beams with perpendicular to the beam axis orientated layers.

4 Theoretical part of the work

The previously presented practical work demonstrates, that CLT and DLT provides an opportunity for beams loaded in bending and by tensile stresses perpendicular to the beam axis direction compared to solid wood or glulam beams. In contrast to CLT and DLT beams, glulam or solid wood beams loaded in bending and by tensile stresses perpendicular to the beam axis direction are prone to splitting. Against it, the reinforcing layers in CLT and DLT beams prevent splitting. Furthermore, the test results show that DLT beams are better than CLT beams in terms of the bending stiffness. Obviously, diagonal orientated reinforcing layers improve the bending stiffness. However, previous investigations were done on a limited number of beams. Due to this limitation, it was not possible to investigate all parameters which assess the load-displacement behaviour of beams loaded in bending. For example, the number of boards and the board dimensions were not investigated so far. The test results are as well not sufficient to determine the stress distribution in this highly complicated CLT or DLT element loaded in bending. To consider all these parameters either more tests are necessary or a finite element analysis is needed.

The following theoretical work is divided into two parts. First, different finite element models for CLT and DLT beams loaded in bending are investigated. A simplified beam model, a plane model and a solid model are presented. These different finite element models are used to demonstrate the dependence of different parameters on the load-displacement behaviour of CLT and DLT beams. Parameters which don't affect the load-displacement behaviour can be neglected. The simplified beam and plane model are used to determine all parameters, which affect the load-displacement behaviour. Based on these results, a solid model is presented.

Using the finite element solid model, which hits the reality best, a parameter study was done. Amongst others the number of parallel to the beam axis orientated timber members and the timber member dimensions were investigated. This investigation helps to understand the difference between the different beam types (CLT vs. DLT) and hence to qualify them.

4.1 Finite element analysis for CLT and DLT beams

4.1.1 Beam model

The simplest finite element model for edgewise orientated CLT and DLT elements loaded in bending is a 2-dimensional model using beam elements. Thereby, each board in any single layer is represented by a beam element with three degrees of freedom (translation in x and y direction and rotation about the nodal z axis). Usually, the boards in real CLT or DLT elements are glued together only over the wide surfaces. The narrow surfaces between neighbouring boards are not glued together. For this reason, this simplified numerical model with beam elements for each single timber member can be used for edgewise orientated CLT and DLT elements. Beams representing the parallel to the beam axis orientated timber members are connected with beams representing the perpendicular or diagonal to the beam axis orientated timber members to a grid. A similar grid model for CLT elements was used by Blaß and Görlacher [2]. Thereby, Blaß and Görlacher describe the load-displacement behaviour of CLT elements loaded as panels in shear using the grid model.

According to Blaß and Görlacher [2], the load-displacement behaviour of edgewise orientated CLT panels is mainly controlled by the connection between the crosswise orientated boards. Only with a rigid bonding between the crosswise orientated boards, the separate casting CLT panel can act as a full plate. However, with degreasing bonding stiffness between the crosswise orientated boards, the CLT panel shear and bending stiffness degreases.

To determine the bonding stiffness between crosswise orientated boards, torsion tests were done according to their real load-carrying behaviour (left in Fig. 21). Thereby two crosswise orientated single timber members were first glued together and after hardening, they were twisted relatively to each other until failure occurred (right in Fig. 21). These tests were used to determine the bonding stiffness and the torsion strength of the connection.

Fig. 21   Left: Test configuration. Right: Test specimen after failure (Figures taken from [2])

Blaß and Görlacher determined the bonding stiffness between 3 N/mm^3 and 8.6 N/mm^3 with an average value of about 5 N/mm^3. The load-displacement behavior of crosswise bonded timber members twisted relatively to each other is characterized by a linear-elastic behavior followed by a brittle failure. The average torsion strength was determined to 3.6 N/mm^2. The minimum torsion strength was 2.6 N/mm^2. Taking into account all test results, Blaß and Görlacher recommend for the bonding stiffness 3 N/mm^3 and for the torsion strength 2.5 N/mm^2.

The bonding stiffness as defined by Blaß and Görlacher was used for the simplified grid or beam model. Thereby edgewise orientated CLT and DLT beams loaded in bending were investigated. The numerical investigations were done using the commercial finite element software ANSYS 10.

ANSYS 10 beam elements were used to create a single supported grid beam loaded in bending. Beam elements were used for parallel to the beam axis orientated boards and for perpendicular or diagonal orientated boards. Two single loads were applied in the third points of the beam length while the supports were applied at the beam ends. Half of the model, which represents a CLT beam, is displayed left in Fig. 22. Right in Fig. 22, half of the model for a DLT beam is displayed.

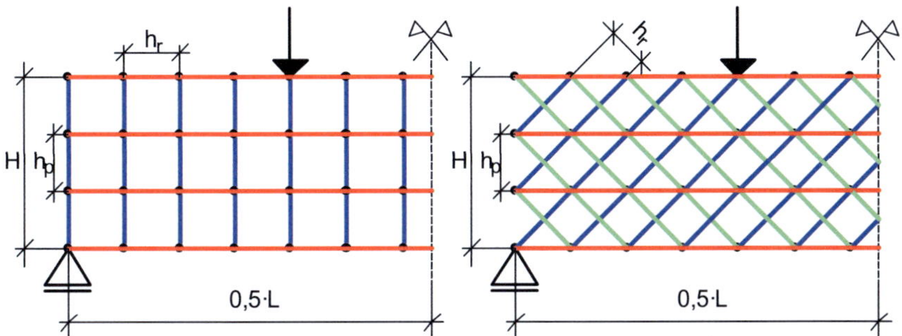

Fig. 22   Grid Model for CLT and DLT beams

The red lines in Fig. 22 represent the neutral axis of the parallel to the beam axis orientated boards. Thereby, their height h_p is equal to the distance between the neutral axes of two side by side and parallel orientated boards. For the CLT beam, the blue lines represent the neutral axis of boards, which are orientated perpendicular to the beam axis direction. Thereby, their board height h_r is equal to the distance between the board axes. In DLT beams, at least two diagonal layers are necessary to form a symmetric beam. Thereby, the diagonal layers must be orientated in the opposite direction to each other. Right in Fig. 22 the green and blue lines represent both diagonal layers which are orientated in the opposite direction to each other. Thereby, their height h_r is equal to the distance between their neutral axes.

ANSYS 10 - Combin14 - spring element was used to represent the torsion behaviour of the glue between the crosswise or diagonal orientated boards (Fig. 23). The end nodes of the Combin14 – spring element provide only one degree of freedom. Only rotation or moments can be transferred between the end nodes, while relative displacements or forces can't be. Therefore, the load-displacement behaviour of the Combine14 - spring element is characterised only by the torsion stiffness.

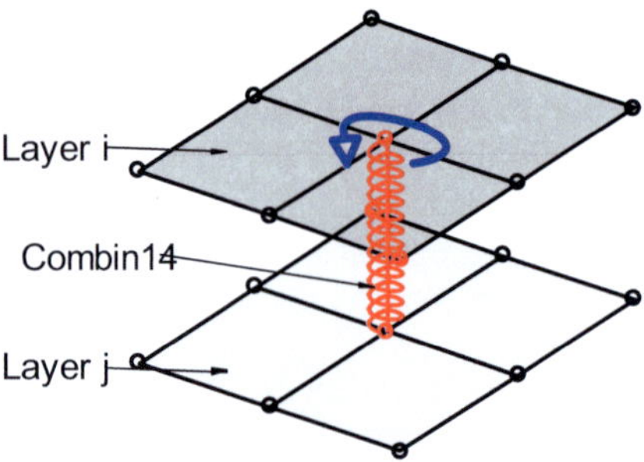

Fig. 23   ANSYS 10 – Combin14 - spring element used for the bonding between the boards

The finite element analysis was done, first to determine the influence of the bonding stiffness on the load-displacement behaviour of CLT and DLT beams and second, to determine the influence of the beam type (CLT vs. DLT) and geometry on the load-displacement behaviour. Therefore following parameters were varied:

Bonding stiffness according to Blaß and Görlacher	$K = 3$ N/mm^3 and 6 N/mm^3
Beam total height	$H = 200$ mm to 1000 mm
Beam total length	$L = 1000$ mm to 10000 mm
Ration between the thickness of the parallel and the perpendicular layers	$t_p / t_r = 0.5$ to 0.9

For all single timber members a constant modulus of elasticity parallel to the grain direction of $E_0 = 12.800$ N/mm^2 was used. The height of the parallel and perpendicular or diagonal to the beam axis orientated boards was chosen to $h_p = h_r = 100$ mm. The bonding connection area affects the torsion stiffness and hence the stiffness of the spring element. The torsion stiffness for the spring element is to calculate using the polar moment of inertia for the connection area and the bonding stiffness K as defined by Blaß and Görlacher. Therefore, the torsion stiffness K_{ser} for the spring element can be calculated according to equation   (10).

$$K_{ser} = \frac{h_p \cdot h_r^3 + h_p^3 \cdot h_r}{12} \cdot K \tag{10}$$

Taking into account all variations, 900 numerical calculations for each CLT and DLT beam were done. All calculations were done on a simple supported beam with two concentrated loads. Thereby the bending stresses in the parallel to the beam axis orientated boards and in particular the beam deflections were calculated. Using the beam deflections, the effective MOE values neglecting the shear component according to equation (2) were calculated. The effective MOE values, depending on the ration between the beam length L and the beam height H, are displayed in the following diagrams.

The first diagram (Fig. 24) contains numerical results for CLT beams with crosswise orientated layers (90°) and with a bonding stiffness $K = 3$ N/mm^3 as defined by Blaß and Görlacher. For a better overview, only a part of the 900 numerical results is displayed. Fig. 25 contains the diagram with numerical results for CLT beams with crosswise orientated layers (90°) and with a bonding stiffness $K = 6$ N/mm^3.

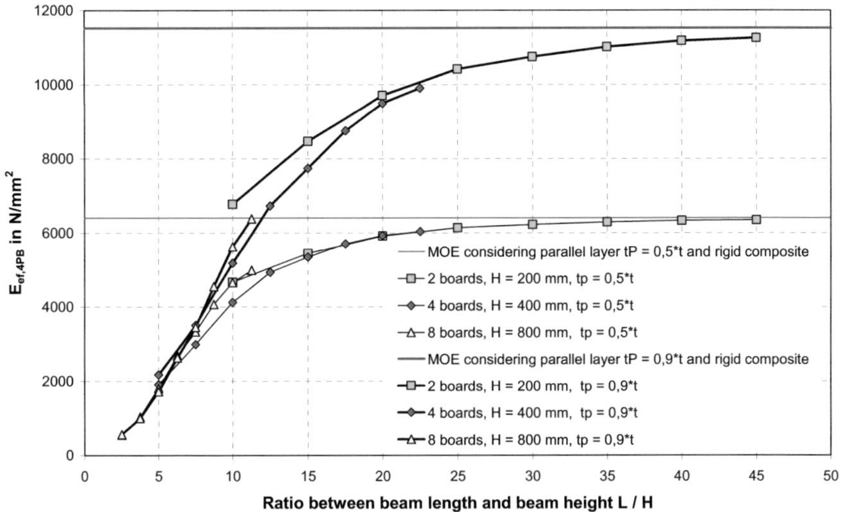

Fig. 24   Effective MOE for CLT beams (90°) with bonding stiffness $K = 3$ N/mm^3

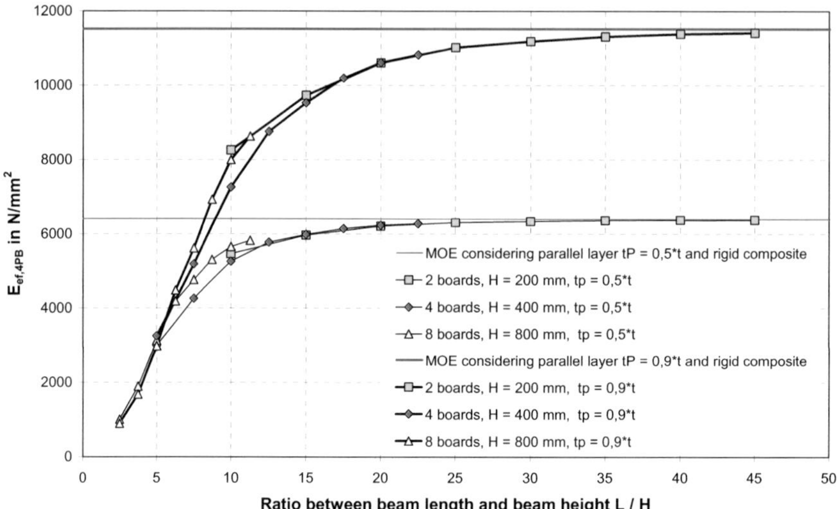

Fig. 25   Effective MOE for CLT beams (90°) with bonding stiffness $K = 6$ N/mm³

Assuming, that the parallel to the beam axis orientated layers with a total thickness t_p act as a full plane, the theoretical MOE for the beam E_{ef} can be calculated using equation (11). Thereby, the stiffness of the perpendicular or diagonal to the beam axis orientated layers with the total thickness t_r is neglected.

$$E_{ef} = E_0 \cdot \frac{t_p}{t_p + t_r} \tag{11}$$

For beams with a ratio between the parallel layers thickness t_p and the total beam thickness t of 0.5, the effective MOE can be calculated to $E_{ef} = 0.5 \cdot E_0 = 6400$ N/mm². For a ratio of 0.9 between the parallel layers thickness t_p and the total beam thickness t, the effective MOE is $E_{ef} = 0.9 \cdot E_0 = 11520$ N/mm². As already mentioned, a MOE value of $E_0 = 12800$ N/mm² was used for all single timber members. Those calculated effective MOE values for beams with $t_p/t = 0.5$ and $t_p/t = 0.9$ are represented in Fig. 24, Fig. 25 and in the following figures by red lines.

Only for large values between the beam length L and the beam height H, the numerical results are nearly similar to the calculated values according to equation (11). Hence, it can be assumed, that the parallel to the beam axis orientated layers act as a full plane. Unfortunately, with degreasing L/H ratio, the effective MOE value degreases. Beams loaded in bending generally are made with length to height ratios between 5 and 20. For CLT beams loaded in bending with L/H-ratios between 5 and 20 and with $K = 3$ N/mm³, the effective MOE can be calculated between 1700 and 9700 N/mm². Due to the low modulus of elasticity and hence due to a low bending stiffness, CLT beams with an L/H-ratio smaller than 20 are not suitable for being loaded in bending.

Even the numerical results considering K = 6 N/mm^3 are not better (Fig. 25). For beams with an L/H-ratio between 5 and 20 the effective stiffness can be calculated between 3.000 and 10600 N/mm^2. A higher bonding stiffness than K = 6 N/mm^3 is hardly to reach. Therefore a higher increase in bending stiffness for CLT beams can not be expected. This results show that CLT elements with crosswise orientated layers even with stiffer bonding are ineffective for beams loaded in bending, in particular for L/H < 20.

Using equal parameters, numerical calculations on DLT beams according to the model right in Fig. 22 with diagonal orientated boards were done. The results for DLT beams with a bonding stiffness of K = 3 N/mm^3 are displayed in Fig. 26. Fig. 27 contains the results for DLT beams with a bonding stiffness of K = 6 N/mm^3.

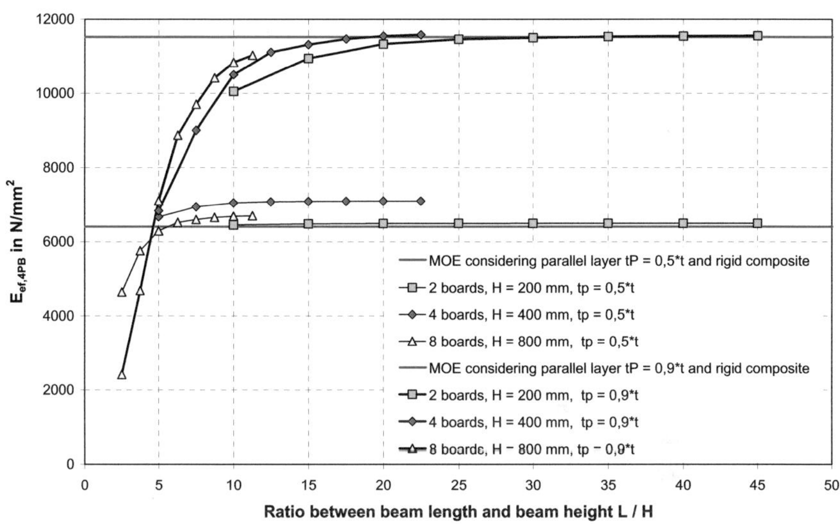

Fig. 26   Effective MOE for DLT beams (45°) with bonding stiffness K = 3 N/mm^3

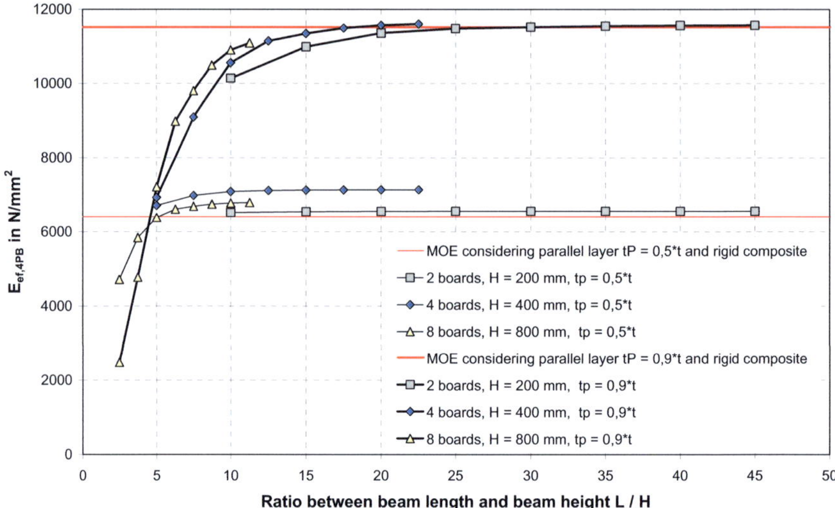

Fig. 27   Effective MOE for DLT beams (45°) with bonding stiffness $K = 6$ N/mm^3

For DLT beams with an L/H-ratio between 5 and 20 the effective MOE can be calculated between 6300 and 11600 N/mm^2. Compared to CLT beams, the effective MOE value and hence the bending stiffness between L/H = 5 and 20 are clearly higher. The numerical calculations with DLT beams show, that the effective MOE and hence the bending stiffness are not affected by the bonding stiffness. For $K = 3$ N/mm^3 and $K = 6$ N/mm^3 the results in Fig. 26 and Fig. 27 are quite similar. Therefore, the bonding stiffness doesn't affect the load-carrying behaviour of DLT beams.

Compared to CLT beams, DLT beams seem to be better in terms of bending stiffness as well for larger L/H-ratios. For longer beams, the numerically determined MOE value is not only similar to the effective MOE value according to equation (11), which already considers the parallel layers as a full plane, but even larger. Obviously, in addition to the parallel to the beam axis orientated layers, the diagonal orientated layers affect the bending stiffness of CLT beams. The highest increase in MOE compared to a full plane was reached for CLT beams with four parallel to the beam axis orientated boards with $t_p/t = 0.5$. Using the finite element analysis, the MOE for those DLT beam was determined to $E_{ef} = 7140$ N/mm^2, which is 12% larger than the corresponding value $E_{ef} = 0.5 \cdot E_0 = 6400$ N/mm^2 according to equation (11).

The presented finite element analysis was done on a simplified model using beam elements. Thereby, beams representing parallel orientated boards were linked together with beams representing perpendicular or diagonal orientated boards using torsion springs. The bonding behaviour between the boards is characterised by the torsion stiffness using torsion springs. The load-displacement behaviour and hence the bending stiffness for suchlike simplified models can be estimated using common calculation models, like for example the shear analogy method (German: Schubanalogieverfahren) or the theory for flexible connected beams. The theory for flexible connected timber beams was first investigated by Möhler in 1956 [3].

Möhler determined this theory to calculate flexible connected beams loaded in bending. Thereby, up to three single timber members can be connected together using nails, dowels, screw and other dowel-type fasteners. The fasteners are orientated along the splice between the timber members to transfer shear loads. Thereby, the fasteners in the splice between the parallel timber members are loaded in the beam axis direction.

Due to the build-up, edgewise orientated CLT and DLT beams loaded in bending can be considered as flexible connected beams. Thereby, the parallel to the beam axis orientated boards represent the continuous beam members. Against it, perpendicular or diagonal to the beam axis orientated boards can be considered as flexibility between the parallel to the beam axis orientated boards. The theory for flexible connected beams considers the flexibility in the splice using a flexibility parameter γ_i. The flexibility parameter γ_i depends on the stiffness K/s along the splice. For lateral loaded dowel-type fasteners, K is the stiffness modulus and s the distance between the dowel-type fasteners in the beam axis direction.

The theory for flexible connected beams according to Möhler can be used as well for CLT and DLT beams loaded in bending. Thereby, first the flexibility parameter γ_i is to determine. For CLT beams loaded in bending, it can be assumed that each perpendicular to the beam axis orientated board acts likewise a lateral loaded dowel-type fastener. The distance between the axes of the perpendicular to the beam axis orientated boards is s. Likewise dowel-type fasteners, perpendicular to the beam axis orientated boards are loaded perpendicular to their axis. Hence, the stiffness K_{CLT} can be calculated assuming a cantilever beam loaded by a single load F at the free end. The span of the cantilever beam is equal to the distance between two parallel to the beam axis orientated boards. The bonding stiffness is represented by two torsion springs at the cantilever beam ends with the torsion stiffness K_R. The stiffness K_{CLT} for each perpendicular to the beam axis orientated board along the splice between the parallel boards depends on the single load F and the lateral deflection u at the cantilever beam ends. The deflection u for a cantilever beam with two torsion springs K_R at the beam ends is to calculate using following equation.

$$u = \frac{F \cdot \ell^3}{12 \cdot E \cdot I} + \frac{F \cdot \ell^2}{2 \cdot K_R} \tag{12}$$

The torsion spring stiffness K_R is the product of the polar moment of inertia and the bonding stiffness K_{ser} as determined by Blaß and Görlacher. The polar moment of inertia depends on the bonding area dimensions, the length a and the width b.

$$K_R = \frac{a \cdot b^3 + a^3 \cdot b}{12} \cdot K_{ser} \tag{13}$$

The stiffness K_{CLT} for each perpendicular to the beam axis orientated board in the splice between the parallel to the beam axis orientated boards can be calculated using following equation.

$$K_{CLT} = \frac{F}{u} = \frac{12}{\ell^2 \cdot \left[\dfrac{\ell}{E \cdot I_r} + \dfrac{6}{K_R} \right]} = \frac{12}{\ell^2 \cdot \left[\dfrac{12 \cdot \ell}{E \cdot b \cdot t_r^3} + \dfrac{72}{a \cdot b^3 + a^3 \cdot b \cdot K_{ser}} \right]} \qquad (14)$$

K_{CLT} and the distance s can be used to calculate the flexibility parameter γ_i and hence the strength and stiffness properties for CLT beams loaded in bending according to [3].

For DLT beams with diagonal to the beam axis orientated beams it can be assumed, that the bonding stiffness does not affect the load-carrying capacity of DLT beam loaded in bending. Numerical results using a simplified finite element analysis confirm this assumption. Therefore, the diagonal to the beam axis orientated beams act more or less as slopes in the splice between the parallel to the beam axis orientated boards. Using this assumption the diagonal orientated boards transfer only tensile loads. Hence, the stiffness K_{DLT} for each diagonal orientated board in the splice between two parallel to the beam axis orientated boards can be calculated according to equation (15). This equation considers the inclination of the diagonal orientated boards.

$$K_{DLT} = \frac{F}{u} = \frac{F}{\varepsilon \cdot \ell} = \frac{F \cdot E}{\sigma \cdot \ell} = \frac{A \cdot E}{\ell} = \frac{b \cdot t_r \cdot E}{\ell} \qquad (15)$$

Using K_{DLT} and the distance s, the flexibility parameter γ_i and hence the strength and stiffness properties for DLT beams loaded in bending can be calculated according to [3].

Using the theory for flexible connected beams according to [3] and equation (14) for CLT beams or equation (15) for DLT beams, the effective MOE depending on the ratio between the beam length L and the beam height H was calculated. To compare numerical and analytical results together, equal parameters as for the finite element analysis were used:

$\ell = h_p = 100$ mm   Height h_p of the parallel boards which is the distance ℓ between the parallel boards axes

$a = h_p = 100$ mm   Height a of the bonding connection area

$b = h_r = 100$ mm   Length b of the bonding connection area

$K_{ser} = 3$ N/mm^3 Bonding stiffness as defined by Blaß and Görlacher

$E_0 = 12.800$ N/mm^2 MOE in grain direction for each single timber member

t_p Total thickness of the parallel to the beam axis orientated layers $t_p = 0.5 \cdot t = 50$ mm and $t_p = 0.9 \cdot t = 90$ mm

Fig. 28 consists of both, numerical results from the finite element analysis and analytical results using equation (14) for CLT beams and equation (15) for DLT beams.

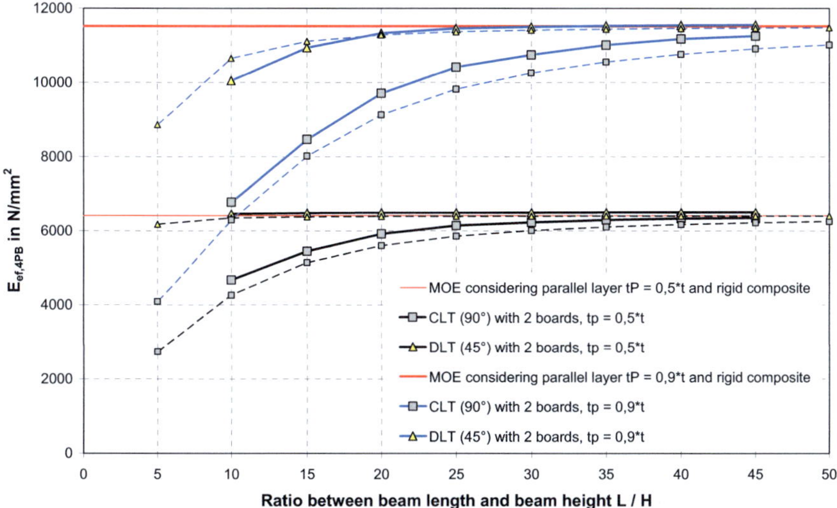

Fig. 28   Effective MOE for DLT beams (45°) and CLT beams (90°) – Comparison between numerical and analytical results

The blue lines represent the results for beams with a total thickness of the parallel to the beam axis orientated layers t_p = 0.9·t. For CLT and DLT beams with t_p = 0.5·t, the results are represented by the black lines. Solid lines represent numerical finite element results while analytical results using the theory for flexible connected beams are represented by the dashed lines. In Fig. 28 only results for beams with two parallel orientated boards are given.

Fig. 28 shows that the analytical results are nearly similar to the numerical finite element results. For this reason, equations (14) and (15), which were used to estimate the flexibility parameter for CLT and DLT beams seem to be realistic. Hence the equations can be used to qualify the beam types and parameters in terms of the bending stiffness of the beams. For a ratio between the cantilever length (distance between the axes of the parallel orientated boards) and the total thickness of the re-inforcing layers ℓ/t_r > 1, the stiffness K_{DLT} for DLT beams is always larger than the stiffness K_{CLT} for CLT beams. This is even valid, when an unlimited bonding stiffness K_{ser} is assumed. For this reason, the effective modulus of elasticity for DLT beams is usually larger than the corresponding value for CLT beams. Hence, in terms of bending stiffness, DLT beams seem to be better than CLT beams.

Both, the finite element analysis and the analytical calculations using a simplified beam model show that DLT beams are quite better than CLT beams in terms of bending stiffness. The results show furthermore, that the bonding stiffness in the range between 3 and 6 N/mm^3 does not affect the bending stiffness of CLT and DLT beams significantly. Nevertheless this simplified model using beam elements and torsion springs is not suitable to determine the load-carrying capacity of CLT and DLT beams. First, this beam model does not consider orthotropic material properties for the single timber members. Second, it is not proved how the friction between neighbored timber members affects the beam performance. And finally, the simplified beam model does not display the real bonding condition, which occurs in CLT and DLT beams. Here, the beams are connected only between two nodes. Against it, in reality, crosswise or diagonal orientated beams are connected over the full surface.

4.1.2 Plane model

The previously presented beam model is not suitable to determine the load-displacement behaviour of CLT and DLT beams loaded in bending. First, orthotropic material properties were neglected using the beam model. Second, the bonding between the single timber members was represented by only one connection node between the torsion spring and the bonding area. Finally, the friction between the single timber members was not considered at all. To improve the finite element analysis a plane model as displayed in Fig. 29 was developed.

Thereby, ANSYS 10 plane elements "Plane182" were used to represent the parallel, perpendicular and diagonal orientated single timber members. These plane elements were fitted with orthotropic material properties for timber. In the splices between neighboured single timber members contact elements using "Contact171" and "Target169" were arranged. These contact elements were placed between the parallel, perpendicular and diagonal to the beam axis orientated timber members.

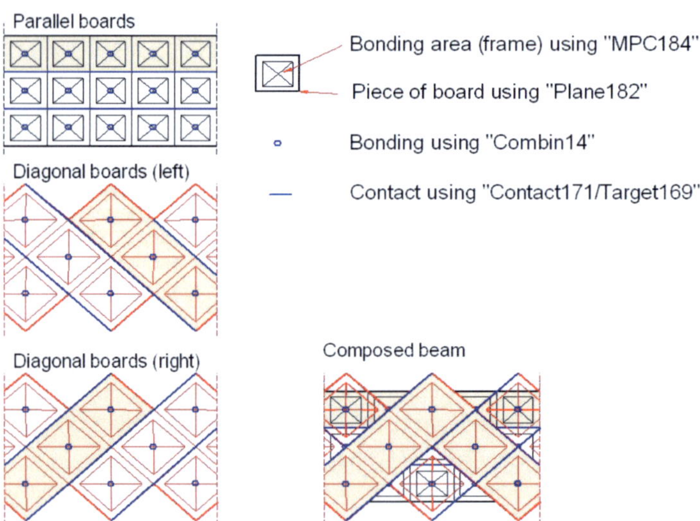

Fig. 29   Plane model considering contact between boards, orthotropic material properties and bonding

Plane182 elements are defined by four nodes with two degrees of freedom at each node. The two degrees of freedom are translation in x and y direction. Unfortunately, rotation is not supported by plane elements. Hence, bonding between timber members as defined by Blaß and Görlacher couldn't be considered using only the Plane182 elements. To solve this problem, an additional supporting frame made out of four beam elements and four diagonal beam elements was arranged in each bonding area between neighboured timber members. This supporting frame was made using rigid beam elements with three degrees of freedom. The frames dimensions were similar to the dimensions of the real bonding area. In contrast to the simplified beam model with only one connection node between two single timber members, the size of the connection area in the plane model was similar to the real connection area.

The four edge nodes of the supporting frame were connected to the nodes of the plane elements. Each bonding surface was equipped with one frame. Hence, each connection had two supporting frames. The diagonal beams in each frame crossed each other in one node, which was arranged in the middle of the frame. Between these middle nodes of two neighboured supporting frames, torsion springs using "Combin14" elements were placed. As for the simplified beam model, the spring elements represented the bonding properties between two timber members.

Although the plane model seems to be better than the beam model, the plane model is characterised by one disadvantage. The plane model can't consider the beam build-up in the out-of-plane direction. For this reason the plane model can' be used to determine, whether a DLT beam with two side by side and diagonal orientated layers is better than a DLT beam with two diagonal layers, which are separated by a parallel to the beam axis orientated layer. This plane model helps merely to qualify the contact and hence the influence of friction between single timber members on the load-carrying behaviour of CLT and DLT beams. To investigate the influence of friction between timber members, 72 numerical calculations using the plane model in Fig. 29 were done. The finite element analysis was done on single supported CLT and DLT beams loaded by two single loads. CLT beams were calculated with two crosswise orientated layers. Against it, for DLT beams three layers are necessary. One layer is orientated in the parallel direction while two layers are orientated in the diagonal and in the opposite direction to each other. Following parameters were varied:

Beam height:	H = 200, 400 and 600 mm
Ratio between beam length and beam height:	L/H = 6, 12, 18 and 24
Friction coefficient:	μ = 0.2, 0.6 and 1.0
Beam type:	CLT and DLT

Following parameters were considered as constant:

MOE in grain direction:	E_0 = 12800 N/mm^2
MOE perpendicular to the grain:	E_{90} = 275 N/mm^2
Shear modulus:	G = 550 N/mm^2
Bonding stiffness:	K = 3 N/mm^3

Ratio between $t_{parallel}$ and $t_{perpendicular}$	$t_P / t_R = 1$
Width of timber members in parallel layers	$h_P = 100$ mm
Width of timber members in perpendicular layers	$h_R = 100$ mm

Using 72 finite element calculations, local and global modulus of elasticity and main stresses were calculated. The global modulus of elasticity consider the shear deflection while the local modulus of elasticity not. To determine the local modulus of elasticity, the relative displacements in the area between the two single loads were determined. Following equation was used to calculate the local modulus of elasticity:

$$E_{ef} = \frac{a \cdot \ell_1^2 \cdot F}{8 \cdot I \cdot \Delta u} \tag{16}$$

With

a   Distance between beam support and neighbouring single load

ℓ_1 Length in the area between single loads

Δu Relative displacement between both measurement points

F   Single load

I   Moment of inertia

Table 9 consist local MOE values E_{ef} for CLT beams. In Table 11 the MOE values for DLT beams are given. To demonstrate the influence of friction, Table 10 and Table 12 are given. In Table 10 and Table 12, based on the E_{ef} for a friction coefficient μ = 0.6, the MOE deviations for μ = 0.2 and μ = 1.0 are given. The differences are very small, even inexistent. Hence, friction does not affect the local modulus of elasticity for CLT and DLT beams in the calculations. Based on these results, contact between narrow surfaces will be neglected for further models.

Table 9 E_{ef} for CLT beams using the plane model

E_{ef} for CLT	H = 200 mm			H = 400 mm			H = 600 mm		
L/H μ	0,2	0,6	1	0,2	0,6	1	0,2	0,6	1
6	1497	**1510**	1524	1408	**1432**	1456	1228	**1256**	1286
12	3873	**3905**	3940	3596	**3677**	3762	3847	**3991**	4140
18	5802	**5822**	5839	5541	**5670**	5800	5833	**5965**	6102
24	6833	**6848**	6941	6553	**6633**	6722	5915	**5980**	6069

Table 10 E_{ef} for CLT beams using the plane model – influence of friction

E_{ef} for CLT	H = 200 mm			H = 400 mm			H = 600 mm		
L/H μ	0,2	0,6	1	0,2	0,6	1	0,2	0,6	1
6	-0,89%	**1510**	0,95%	-1,72%	**1432**	1,68%	-2,31%	**1256**	2,34%
12	-0,81%	**3905**	0,89%	-2,21%	**3677**	2,29%	-3,60%	**3991**	3,73%
18	-0,34%	**5822**	0,30%	-2,28%	**5670**	2,29%	-2,22%	**5965**	2,30%
24	-0,22%	**6848**	1,35%	-1,21%	**6633**	1,33%	-1,09%	**5980**	1,50%

Table 11 E_{ef} for DLT beams using the plane model

E_{ef} for DLT	H = 200 mm			H = 400 mm			H = 600 mm		
L/H μ	0,2	0,6	1	0,2	0,6	1	0,2	0,6	1
6	5431	5476	5517	6343	6432	6514	6729	6782	6824
12	7307	7324	7343	7292	7305	7317	7247	7253	7257
18	7585	7591	7595	7370	7372	7375	7282	7283	7283
24	7741	7743	7744	7402	7402	7403	7293	7293	7293

Table 12 E_{ef} for DLT beams using the plane model – influence of friction

E_{ef} for DLT	H = 200 mm			H = 400 mm			H = 600 mm		
L/H μ	0,2	0,6	1	0,2	0,6	1	0,2	0,6	1
6	-0,81%	5476	0,75%	-1,38%	6432	1,29%	-0,79%	6782	0,61%
12	-0,23%	7324	0,25%	-0,18%	7305	0,16%	-0,08%	7253	0,06%
18	-0,08%	7591	0,06%	-0,04%	7372	0,03%	-0,02%	7283	0,01%
24	-0,02%	7743	0,02%	0,00%	7402	0,01%	0,00%	7293	0,00%

Following results present the difference between CLT and DLT beams. Fig. 30 consists numerical results for a friction coefficient μ = 0.6. The relationship between local MOE values and the ratio between the beam length and the beam height is shown. The solid lines represent the results for DLT beams, while the mashed lines represent the results for CLT beams. As already presented (see beam model), the MOE values for DLT beams are higher than the corresponding values for CLT beams. For this reason, DLT beams loaded in bending are more effective in terms of bending stiffness than comparable CLT beams.

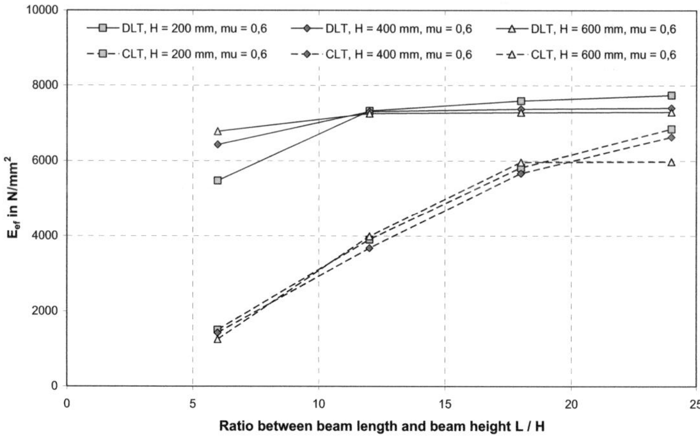

Fig. 30   Effective MOE E_{ef} for CLT and DLT elements (plane model with μ = 0.6)

4.1.3 Solid model

A simplified beam model and a modified plane model were presented. Both models have some restrictions. Hence, they are not suitable to determine the load-displacement behaviour of CLT and DLT beams loaded in bending. The beam model doesn't consider orthotropic material properties and represents the bonding between two single timber members using one node. Although, the plane model considers orthotropic material properties and a bonding connection area with realistic dimensions, the plane model cannot be used to determine the influence of the beam build-up in the out-of-plane direction. To evade these restrictions, a solid model was developed. The solid model was made using ANSYS 10 solid elements "Solid95". The element "Solid95" is characterised by 20 nodes with three degrees of freedom. Among others, orthotropic material properties were considered. In contrast to the previously presented models, the solid model considers all three dimensional directions. Hence, the real beam build-up can be considered. For this reason, CLT beams with five plies and in particular DLT beams with four and five plies were investigated. All numerical investigations were done considering the same beam build-up as for the test specimens.

The previously presented numerical results show, that the bonding stiffness in the range as defined by Blaß and Görlacher, doesn't affect the load-displacement behaviour of CLT and DLT beams loaded in bending. Therefore, the bonding stiffness as defined by Blaß and Görlacher is neglected in the solid model. Due to the fact, that the solid model considers as well the beam build-up in the out-of-plane direction, the contact between single timber members can be presented by simple connection. Thereby, the nodes in the bonding connection area between two neighboured single timber members are glued together. This special fusion between two single timber members around the bonding connection area was realised using contact elements ("Contact174" and "Target170"). To prevent sliding between the nodes, "keyoption 12" was set to 5. For KO 12 = 5, contact and target elements are always bonded together.

In the splices between the narrow surfaces no contact elements were applied. Using the plane model it was demonstrated, that friction between the narrow surfaces of neighboured boards does not affect the load-displacement behaviour of CLT and DLT beams. The friction between the narrow surfaces of neighboured boards should be neglected anyway, because it's difficult to control. In reality, the gap thickness varies due to changing moisture content. In case of swelling, gaps become smaller and in case of shrinkage they become larger. Gaps with 6 mm thickness are usual in larger CLT and DLT elements. Considering a gap width of 6 mm, contact between the narrow surfaces of neighboured boards cannot be transferred at small deflections anyway.

First, a finite element analysis using the solid model was done to compare the numerical results with the test results. Using the finite element analysis, CLT and DLT beams were analysed, which are completely similar to the tested specimens in terms of geometry and material properties. Numerical investigations were done on short (series 1-1) and long (series 1-2) CLT beams with five plies. Furthermore, short (series 2-1) and long (series 2-3) DLT beams with five plies as well as short (series 2-2) and long (series 2-4) DLT beams with four plies were investigated. The geometry for all beams is summarised in chapter 3.4.1 and 3.4.2. The beam build-up is given in Fig. 47 to Fig. 52. All beams were loaded by the average ultimate load from previous tests.

Table 3 consist the ultimate loads for short beams while Table 4 deals with the ultimate loads for long beams. The finite element analysis was done considering a modulus of elasticity in grain direction of 11900 N/mm^2. This MOE corresponds to the average MOE $E_{50\%}$, which was determined for the single timber members used for the parallel layers (Table 2). Further stiffness parameters, like the MOE perpendicular to grain and the shear modulus were not previously determined. Based on the DIN 1052:2004-08, the MOE perpendicular to the grain direction $E_{90°}$ was assumed as $E_{90°} = E_{0°} / 30 = 400$ N/mm^2. The shear modulus G was assumed as G = 750 N/mm^2 while the rolling shear modulus G_{RS} was assumed as $G_{RS} = G / 10 = 75$ N/mm^2. The chosen ratio between E_0 and E_{90} and between G and G_{RS} corresponds to the ratio for timber according to DIN 1052:2004-08.

The investigated systems are displayed in Fig. 96 (specimen 1-1), Fig. 99 (specimen 2-1), Fig. 102 (specimen 2-2), Fig. 105 (specimen 1-2), Fig. 108 (specimen 2-3) and Fig. 111 (specimen 2-4). Those figures present full CLT and DLT beam systems with supports and loading plates. Parallel and perpendicular or diagonal to the beam axis orientated layers are displayed separately. Fig. 96 to Fig. 113 show beams under loading and the axial stresses distribution. In addition, the lateral deformation in the out-of-plane direction (z-axis) is displayed. The beam deformation in the out-of-plane direction was scaled with factor five for short beams and with factor three for long beams to improve the view. Hence, it can be noticed, that DLT beams with diagonal to the beam axis orientated layers are prone to twist. Both diagonal layers are orientated in the opposite direction and with an eccentricity to the beam axis. Due to this orientation, DLT beams are loaded in torsion which leads to twisting. The torsion moment increases with increasing shear load and with increasing eccentricity of the diagonal orientated layers. Therefore, DLT beams should be produced with diagonal orientated layers located as close as possible to the beam axis and in between the parallel layers. Then, torsion moments can be reduced and beam twisting can be prevented. Outside on the beam surface orientated diagonal layers should be avoided. Although the investigated DLT beams are prone to twist, the tested and numerically calculated beams never failed in buckling. Compared to the beam size, the deflection in the out-of-plane direction is too small.

To qualify the different beam types, stresses and stiffness properties were determined. For all six beam types, normal stresses in the parallel to the beam axis orientated layers along path 1 were determined. Shear stresses for all six beam types were determined along path 2. It can be assumed, that the maximum normal stresses in the parallel to the beam axis orientated layers occur along path 1 while the maximum shear stresses in the parallel orientated layers occur along path 2. The definition of the different paths considering the different beam types is given in Fig. 31 and Fig. 32.

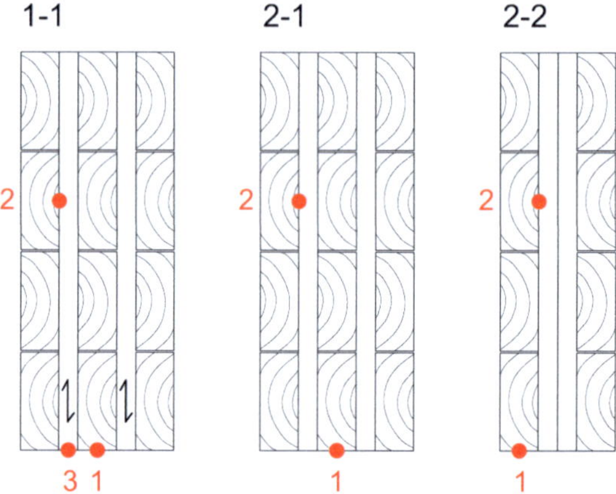

Fig. 31 Path definition for short beams (path 1 for σ_x, path 2 for τ_{xy}, path 3 for σ_y)

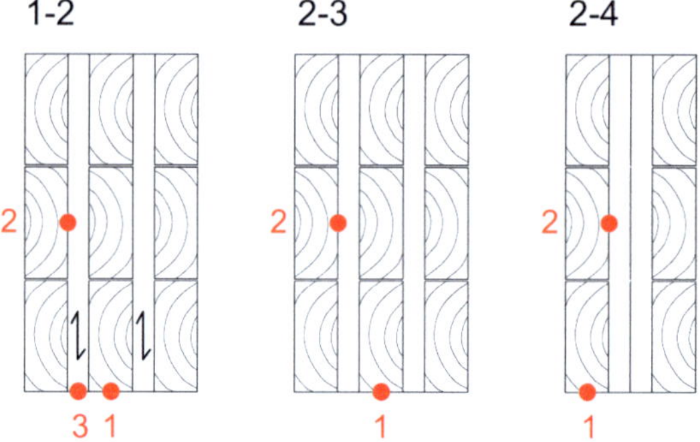

Fig. 32 Path definition for long beams (path 1 for σ_x, path 2 for τ_{xy}, path 3 for σ_y)

In addition, (tensile) stresses perpendicular to the grain direction in the perpendicular to the CLT beam axis orientated inner layers along path 3 were determined. Due to the bending of the CLT beams, the top beam surface is loaded in compression while the bottom beam surface is loaded in tension. Therefore, perpendicular to the beam axis orientated boards are loaded in tension perpendicular to their grain direction. This investigation was done, because timber is very week in tension perpendicular to the grain direction. However, tensile stresses perpendicular to the grain direction in the diagonal to the beam axis orientated boards in DLT beams were not determined. Compared to the perpendicular to the beam axis orientated boards in CLT beams, the diagonal orientated boards in DLT beams are loaded by much smaller tensile stresses perpendicular to the grain direction due to the diagonal orientation.

Normal stresses in the parallel layer along path 1 are presented for short beams in Fig. 33 and for long beams in Fig. 34. Although the distribution and the maximum values seem to be equal, they are quite different. The maximum bending stress for the short CLT beam was calculated to 46.3 N/mm^2, while the maximum bending stress for the short DLT beam with five plies was calculated to 45.7 N/mm^2 (-1%). Although the bending stresses and the cross sections are quite similar, the short DLT beam with five plies is better than the short CLT beam. The short CLT beam was loaded by 270 kN, while the short DLT beam with five plies was loaded by 301 kN (+12%). Even better is the short DLT beam with four plies. Assuming a total loading of 229 kN, the maximum bending stress was calculated to 47.8 N/mm^2 which is quite similar to the maximum stresses for the short CLT beam. However, the short DLT beam with four plies was loaded by 229 kN which corresponds to an effective loading of 105/70·229 kN = 344 kN considering only two parallel layers with a total thickness of 70 mm for the DLT beam with four plies and three parallel layers with a total thickness of 105 mm for the CLT beam with five plies. For this reason, the performance of the investigated DLT beam with four layers is 27% higher compared to the investigated short CLT beam.

Similar results were made for long beams (Fig. 34). The long CLT beam was loaded by 91.8 kN as determined in tests. Thereby, the maximum bending stress was determined to 47.6 N/mm^2. The long DLT beam with five plies was loaded by 87.1 kN which is 5.1% smaller than the corresponding load for the long CLT beam. However, the maximal bending stress was determined to 41.6 N/mm^2 which is even 12.6% smaller than the maximum bending stress value for the long CLT beam. Therefore, the performance of the long DLT beam with five plies is about 7.5% better compared to the long CLT beam. Even better are the results for the long DLT beam with four plies. The long DLT beam with four plies was loaded by 58.3 kN or being transferred to a thickness of totally 105 mm for the parallel layers, by 105/70·58.3 kN = 87.5 kN. Compared to the long CLT beam, the effective loading is quite similar (-0.2%). However, the calculated maximum bending stress for the long DLT beam with four plies was 40.4 N/mm^2 which is 15.1% smaller than the corresponding value for the long CLT beam. Therefore, the performance of the long DLT beam with four plies is about 15% better compared to the investigated long CLT beam.

Taking into account the bending stresses, DLT beams with four plies seem to be better than DLT beams.

The main results are as well summarized in Table 13. Column three consist loads, which were used for the finite element analysis. These values are equal to the corresponding test results. The maximum bending stresses along path 1 for the different beams, which correspond to the loads in column three, are given in column four.

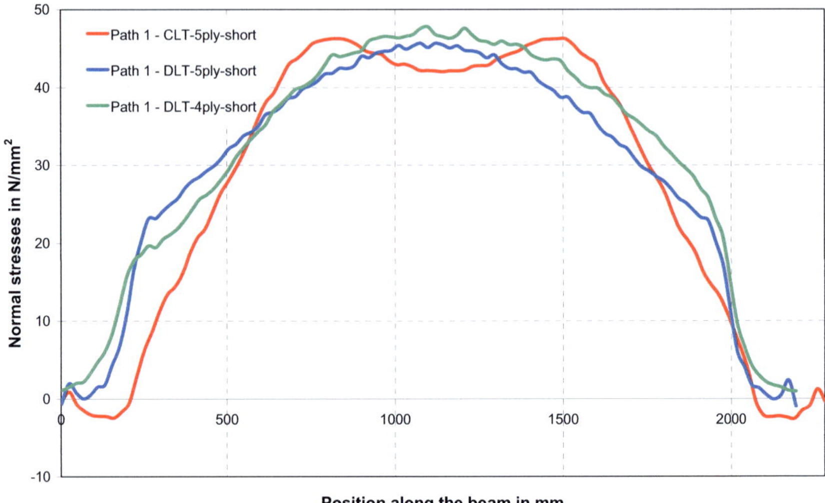

Fig. 33   Distribution of normal stresses σ_x along path 1 for short beams

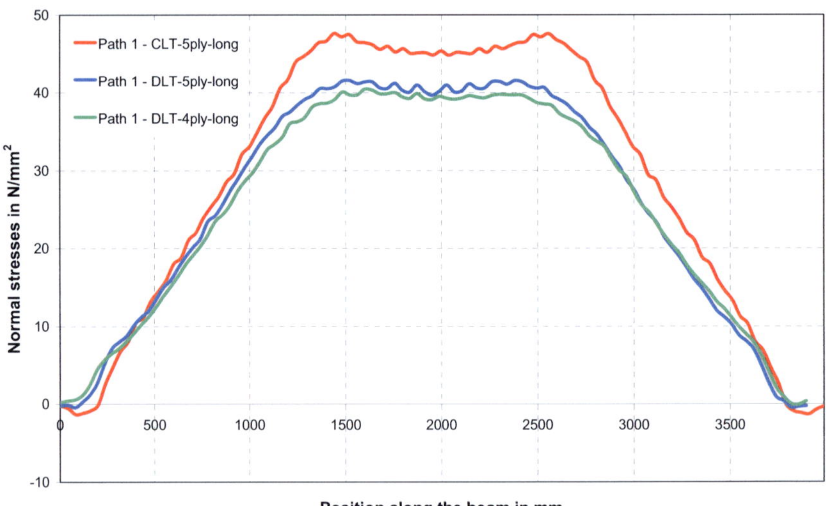

Fig. 34   Distribution of normal stresses σ_x along path 1 for long beams

Shear stresses distribution along path 2 for short beams is presented in Fig. 35. Fig. 36 contains the shear stresses distribution along path 2 for long beams. The maximum values for shear stresses along path 2 are given in column five in Table 13. The determined shear stresses for CLT and DLT beams with five plies are quite similar. Against it, for short and long DLT beams with four layers, the determined shear stresses are lower than the corresponding values for the other beam configurations. Using the beam theory and taking into account only the parallel layers, the maximum shear stresses for the short CLT beam, loaded by 270 kN, can be calculated to $\tau_{xy,net,par}$ = 5.36 N/mm^2. Against it, according to the finite element analysis result, the maximum shear stress is τ_{xy} = 5.82 N/mm^2 (+8.6%). The maximum shear stress for the short DLT beam with five plies loaded by 301 kN can be calculated to $\tau_{xy,net,par}$ = 5.98 N/mm^2 according to the beam theory. Against it, the finite element analysis provides a maximum shear stress, which is τ_{xy} = 5.12 N/mm^2 (-13%). The short DLT beam with four plies is even better. The maximum shear tress for the short DLT beam with four plies loaded by 229 kN can be calculated to $\tau_{xy,net,par}$ = 6.82 N/mm^2 considering the beam theory. Thereby, the maximum shear stress in the finite element analysis was determined to τ_{xy} = 3.70 N/mm^2, which is 46% smaller than the calculated shear stress according to the beam theory. Using the finite element analysis, the maximum shear stress for the short CLT beam is even 8.6% higher than the corresponding value according to the beam theory. For the short DLT beam with five plies, the numerically determined shear stress is 13% smaller than the corresponding value according to the beam theory. Against it, the maximum shear stress for the short DLT beam with four plies is about 46% smaller than the corresponding value according to the beam theory. Hence, the investigated short DLT beams with four plies provide the best shear performance.

In terms of shear stresses, long beams behave similar to the short beams. For the long CLT beam loaded by 91.8 kN, the shear stress can be calculated to $\tau_{xy,net,par}$ = 2.43 N/mm^2 according to the beam theory while the maximum shear stress using the finite element analysis was determined to τ_{xy} = 2.84 N/mm^2 (+17%). For the long DLT beam with five plies loaded by 87.1 kN the shear stress can be calculated to $\tau_{xy,net,par}$ = 2.30 N/mm^2 according to the beam theory while the maximum shear stress using the finite element analysis was determined to τ_{xy} = 2.19 N/mm^2 (-4.8%). The long DLT beam with four plies is even better. The long DLT beam with four plies loaded by 58.3 kN, is loaded by a maximum shear stress $\tau_{xy,net,par}$ = 2.31 N/mm^2 according to the beam theory. Against it, using the finite element analysis, the maximum shear stress was determined to τ_{xy} = 1.35 N/mm^2, which is 42% smaller than the shear stress according to the beam theory. Likewise for short beams, the shear performance for DLT beams with four plies is the best while the shear performance for CLT beams is the worst.

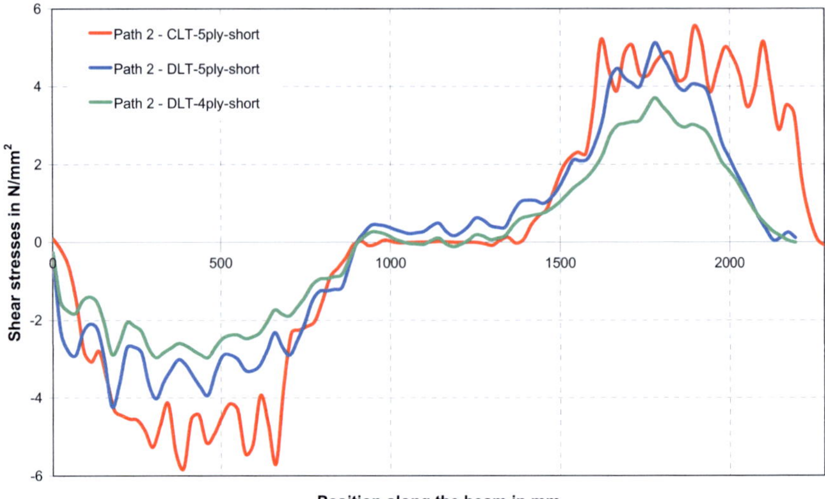

Fig. 35　　Distribution of shear stresses τ_{xy} along path 2 for short beams

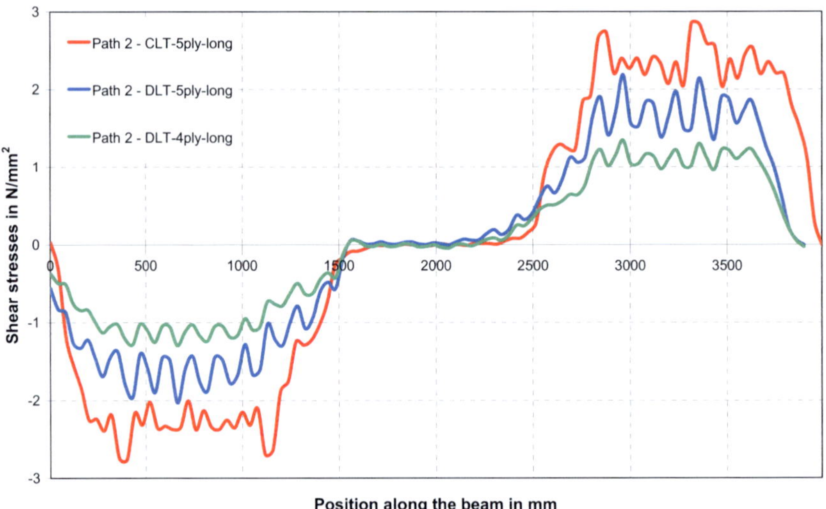

Fig. 36　　Distribution of shear stresses τ_{xy} along path 2 for long beams

Finally, tensile stresses perpendicular to the grain direction in the inner layers of CLT beams were investigated. The distribution for the investigated short CLT beam is displayed in Fig. 37. Fig. 38 contains the stresses distribution perpendicular to the grain direction for the long CLT beam. The maximum values are given in Table 13 in column six. The larger the ratio between the beam deflection and the beam length, the higher the tensile stresses at the bottom surface of the beams. The maximum tensile stresses perpendicular to the grain direction were determined to $\sigma_{y,max}$ = 1,29 N/mm^2 for the short CLT beam and $\sigma_{y,max}$ = 1,28 N/mm^2 for the long CLT beam. Compared to the very low tensile strength perpendicular to the grain direction for timber, the stresses reached almost the tensile strength. Taking into account the DIN 1052:2004-08, the characteristic tensile strength perpendicular to the grain direction is $f_{t,90,k}$ = 0.4 N/mm^2. However, the average value for the tensile strength perpendicular to the grain direction is usually significantly higher and reaches sometimes values around 1 N/mm^2 and even more. For this reason, no cracks or splitting was observed in the test specimens. Furthermore, cracks and splitting in the perpendicular to the beam axis orientated boards were obviously prevented by the parallel to the beam axis orientated boards. Nevertheless, the results show, that tensile stresses perpendicular to the grain direction in particular in the perpendicular to the beam axis orientated boards in CLT beams occur and can't be neglected.

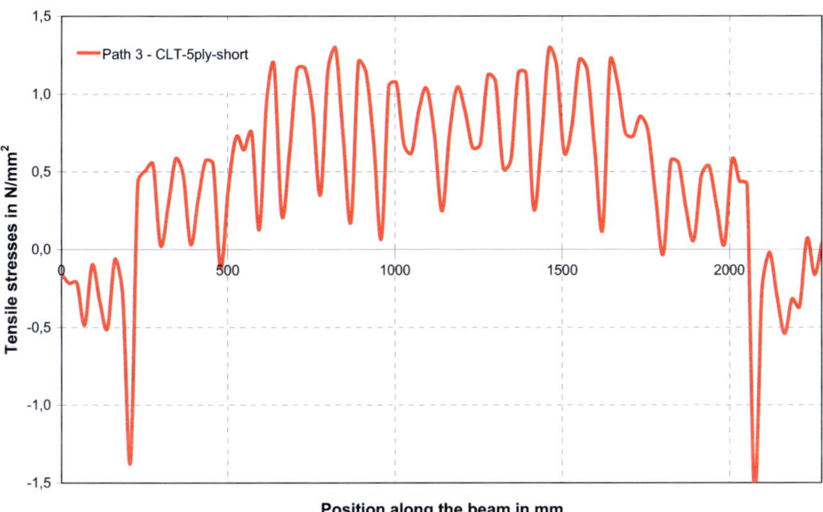

Fig. 37   Distribution of stresses perpendicular to beam axis σ_y along path 3 for short beams

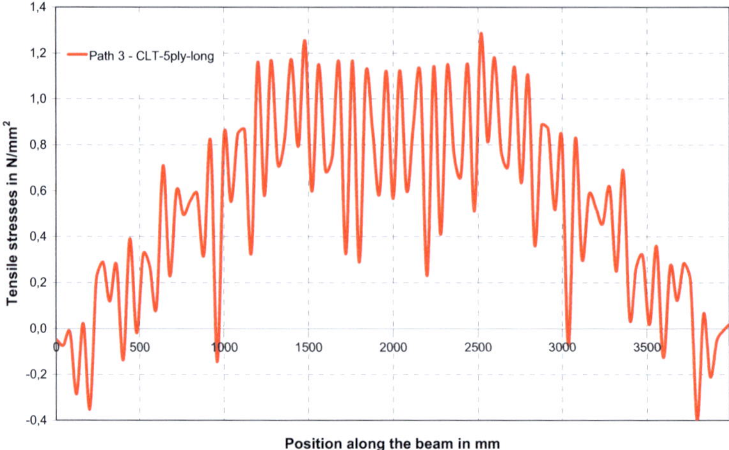

Fig. 38   Distribution of stresses perpendicular to beam axis σ_y along path 3 for long beams

In the following Table 13 all numerical results are summarised. The values in column one to six were already discussed. Furthermore, MOE values for the investigated CLT and DLT beams were determined. In column seven the local MOE is given. This MOE was determined using the whole beam thickness and equation (16). In column eight the average test results are given. These values were determined considering the net timber thickness, which corresponds to the total thickness of the parallel layers. The ratio between the MOE values, determined in the finite element analysis, and the corresponding test result is displayed in column nine. For short beams the ration between E and E_0 is around 90%. Against it, the ratio for long beams is around 100%. Taking into account all results, in particular those for the MOE values, the investigated finite element solid model is suitable to determine the load-displacement behaviour of CLT and DLT beams loaded in bending.

Table 13   Summarisation of the finite element analysis

		F_{tot}	$\sigma_{x,max}$	$\tau_{xy,max}$	$\sigma_{y,max}$	E	E_0	E / E_0
		[kN]	[N/mm²]	[N/mm²]	[N/mm²]	[N/mm²]	[N/mm²]	[-]
1-1	CLT-5ply-short	270	46,3	5,82	1,29	7214	7782	93%
2-1	DLT-5ply-short	301	45,7	5,12	-	6370	7769	82%
2-2	DLT-4ply-short	229	47,8	3,70	-	6756	7170	94%
1-2	CLT-5ply-long	91,8	47,6	2,84	1,28	9240	9206	100%
2-3	DLT-5ply-long	87,1	41,6	2,19	-	9319	9075	103%
2-4	DLT-4ply-long	58,3	40,4	1,35	-	8658	8672	100%

The finite element analysis results from the solid model are nearly similar to the test results. Therefore, the solid model fits the real load-displacement behaviour of CLT and DLT beams loaded in bending the best. In the second step of the theoretical part of the work, a parameter study was done using the solid model. Thereby, short and long CLT beams with five plies and short and long DLT beams with four and five plies were calculated varying following parameters:

Number of parallel boards in each parallel layer: n_p = 2, 3, 4, 5 and 6
Width of the perpendicular or diagonal orientated board
compared to the width of the parallel orientated board: h_r = 0.5·h_p, 1.0·h_p, 1.5·h_p
Total thickness of all perpendicular or diagonal layers: t_r = 20, 35 and 50 mm

Considering all parameters, 45 numerical calculations for each beam type were done. For six different beam types, altogether 270 numerical calculations were done. Other parameters were used as taken for the tested specimens. Hence, short beams were 360 mm in the height and 2300 mm in the length. The thickness of the parallel layers was chosen to 35 mm which is similar to the used boards. Long beams were 270 mm in the height and 4000 mm in the length. The thickness of the parallel layers was chosen as well to 35 mm. The loads and supports were applied at the same position likewise for the tests.

The stiffness parameters (MOE and shear modulus) were equal to those which were already used for the first investigations. The modulus of elasticity in grain direction was used as 11900 N/mm^2. This MOE corresponds to the average MOE $E_{50\%}$, which was determined for the single timber members used for the parallel layers (Table 2). Further stiffness parameters, like the MOE perpendicular to the grain direction and the shear modulus were not determined. Therefore, the MOE perpendicular to the grain direction $E_{90°}$ was assumed as $E_{90°}$ = $E_{0°}$ / 30 = 400 N/mm^2. The shear modulus G was assumed as G = 750 N/mm^2 while the rolling shear modulus G_{RS} was assumed as G_{RS} = G / 10 = 75 N/mm^2. The ratio between E_0 and E_{90} as well as the ratio between G and G_{RS} corresponds to the ratio for timber according to DIN 1052:2004-08.

First, the local modulus of elasticity neglecting the shear influence according to equation (16) was determined for all 270 beams. Detailed results are given in Fig. 114 to Fig. 119. Each figure contains results for a single beam type. The local MOE was determined considering the total beam thickness, including parallel, perpendicular and diagonal to the beam axis orientated layers. The results are given in dependence of the number of parallel boards per single layer (x-axis). Red lines represent beams, which were calculated with a total thickness of the perpendicular or diagonal layers of t_r = 50 mm. Beams with a total thickness of the perpendicular or diagonal layers t_r = 35 mm are represented by the blue lines. And finally, black lines represent beams with t_r = 20 mm. The total thickness of the parallel layers was t_p = 105 mm for beams with five plies and t_p = 70 mm for beams with four plies. The ratio between the width of the inner layers h_r and the width of the parallel layers h_p is varied by the different lines, too.

Subsequent, only results for beams with a thickness ratio of $t_p/t = 75\%$ (78%) are summarised in two diagrams. Fig. 39 contains the results for short beams, while results for long beams are summarised in Fig. 40. These diagrams are used to qualify the different beam types in terms of the effective modulus of elasticity.

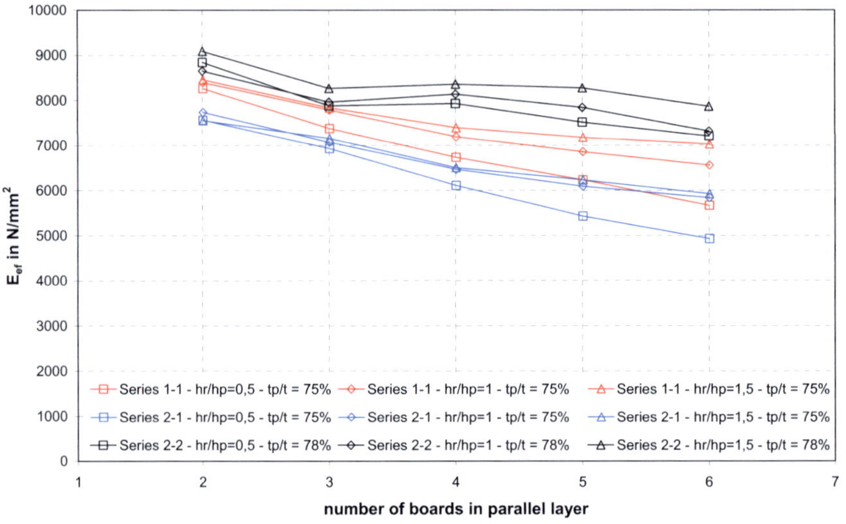

Fig. 39 E_{ef} for short beams with $t_p/t = 75\%$ (78%)

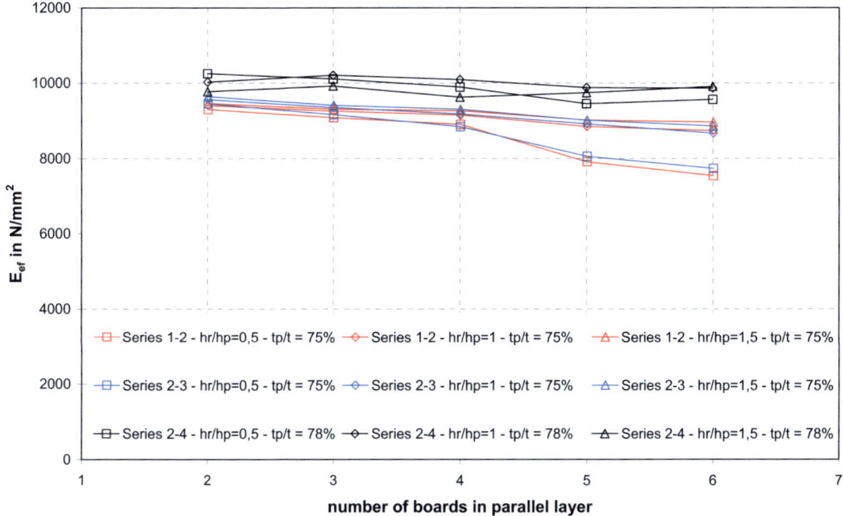

Fig. 40 E_{ef} for long beams with $t_p/t = 75\%$ (78%)

Both diagrams show clearly, that DLT beams with four plies (series 2-2 and 2-4) are better than the other two beam types. Compared to CLT (series 1-1 and 1-2) and DLT beams with five plies (series 2-1 and 2-3), the MOE values for DLT beams with four plies are always higher. Thereby, the thickness ratio for beams with four and five plies is quite similar. Fig. 39 and Fig. 40 contain results for beams with $t_p/t = 75\%$ and 78%. The ratio $t_p/t = 75\%$ corresponds to beams with five plies (t = 140 mm, t_p = 105 mm, t_r = 35 mm), while the ratio $t_p/t = 78\%$ corresponds to beams with four plies (t = 90 mm, t_p = 70 mm, t_r = 20 mm). In spite of the small difference between t_p/t for beams with five and four plies, the results are pretty comparable. As already mentioned, two side by side and in the opposite direction orientated diagonal layers increase the modulus of elasticity and hence the bending stiffness for the beams. Against it, diagonal orientated layers, which are separated by a parallel layer (DLT beam with five plies), are not able to increase the bending stiffness. Obviously, due to their separation by at least one parallel layer, they cannot be considered as a full plane.

Furthermore, the results present an effect, which was previously mentioned using the theory for flexible connected beams. With increasing number of parallel to the beam axis orientated boards, the beam MOE degreases. This undesirable effect is large for CLT and DLT beams with five plies and is quite small for DLT beams with four plies. In particular, long DLT beams with four plies aren't affected by this undesirable effect. One more factor, which affects the beam MOE, is the width h_r/h_p of the perpendicular or diagonal to the beam axis orientated boards. With degreasing width h_r the MOE degreases. Considering all results it can be concluded, that with increasing amount of single timber members used for CLT or DLT beams, the beam MOE and the bending stiffness degreases.

Fig. 39 and Fig. 40 contain the MOE values E_{ef}, which consider the whole timber thickness. However, of larger interest is the ratio between the MOE values for CLT and DLT beams and the MOE values for geometrical identical solid beams. Subsequently, the MOE values $E_{ef,par}$ for CLT and DLT beams according to equation (17) were determined.

$$E_{i,par} = E_i \cdot \frac{t_{p,tot}}{t_{p,tot} + t_{r,tot}} \qquad\qquad (17)$$

These values were compared to the MOE values $E_{net,par}$ for geometrical identical solid beams. The whole results, which show the relationship between the ratio $E_{ef,par}$ / $E_{net,par}$ and the number of parallel boards, are given in Fig. 120 to Fig. 125. Each figure contains the results for each single beam type. Both MOE values, $E_{ef,par}$ and $E_{net,par}$, consider only the thickness of the parallel to the beam axis orientated layers. The MOE $E_{net,par}$ for geometrical identical solid beams was calculated using equation (17) and considering the average MOE for the single timber members (E_0 = 11900 N/mm^2). The ratio between $E_{ef,par}$ and $E_{net,par}$ helps to determine the beam quality in terms of bending stiffness. As long as the ratio equals 100%, the parallel layers can be considered as a full plane. Is the ration between $E_{ef,par}$ and $E_{net,par}$ smaller than 100%, the parallel layers cannot be considered as a full plane. In this case, the flexibility between the parallel layers is to low. Against it, a ration between $E_{ef,par}$ and $E_{net,par}$ greater than 100% occur, when the parallel to the beam axis orientated layers act as a full plane and in addition, the perpendicular or diagonal layers act as an additional plane.

Subsequently, the different results are discussed. For better understanding, only results for beams with a thickness ratio of t_p/t = 75% (78%) are taken into account. Fig. 41 contains results for short beams, while results for long beams are given in Fig. 42. The remaining results are displayed in the attachment in Fig. 120 to Fig. 125.

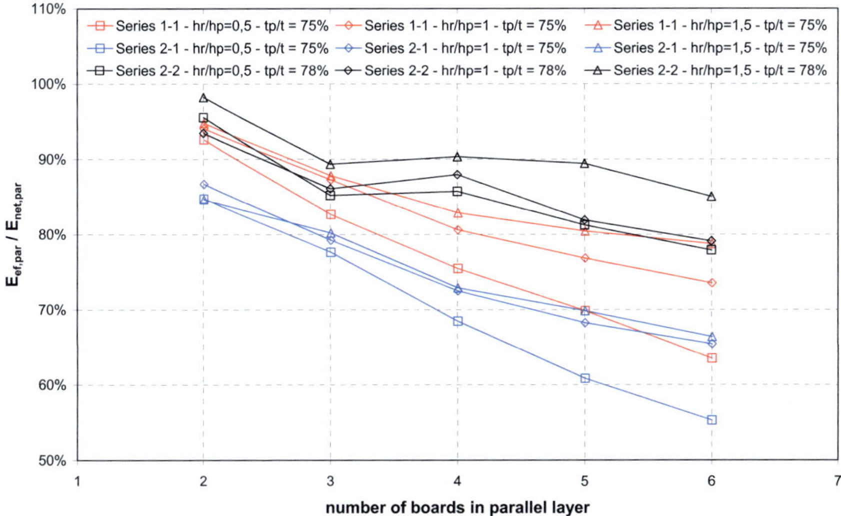

Fig. 41   Ratio between E_{ef} and E_{net} for short beams with t_p/t = 75% (78%)

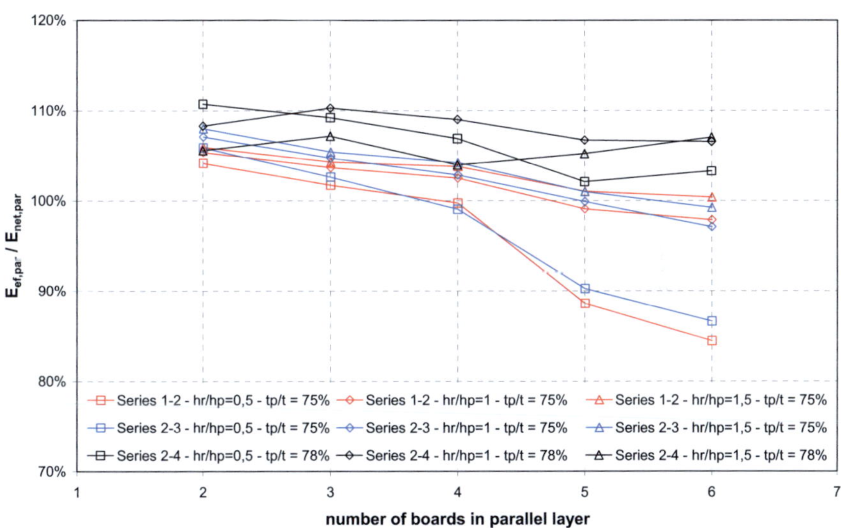

Fig. 42   Ratio between E_{ef} and E_{net} for long beams with t_p/t = 75% (78%)

Taking into account both, short and long beams, the ratio between $E_{ef,par}$ and $E_{net,par}$ for DLT beams with four plies is higher than the corresponding ratio for CLT and DLT beams with five plies. As displayed in Fig. 41, the ratio $E_{ef,par}$ / $E_{net,par}$ is for all investigated short beams smaller than 100% and degreases with increasing number of parallel boards. Usually, the load-carrying capacity for short solid beams isn't governed by bending stiffness and by deflection. However, due to the low ratio $E_{ef,par}/E_{net,par}$, short CLT beams in particular with numerous parallel boards should be checked even for bending stiffness and for deflection. The smallest value between $E_{ef,par}$ and $E_{net,par}$ was about 55%. For this reason, CLT beams with many parallel boards are prone to large deflection which can govern their load-carrying capacity. Against it, the degreasing in MOE for short DLT beams with four layers is much smaller.

Long beams are prone to large deflections. Hence, the bending stiffness can govern their load-carrying capacity. Therefore, long beams should be detailed to minimise their deflection and to maximise their bending stiffness. According to Fig. 42, DLT beams with side by side orientated diagonal layers should be used for longer beams. For the investigated DLT beams with four plies, the ratio $E_{ef,par}$ / $E_{net,par}$ is always larger than 100%. Obviously, the two side by side and in the opposite direction orientated diagonal layers act as an additional full panel, which increase the effective MOE and the bending stiffness of the beam.

Against it, the MOE for CLT and DLT beams with five plies is lower and degreases with increasing number of parallel to the beam axis orientated boards or/and with degreasing width h_r of the reinforcing layers. CLT beams loaded in bending should be detailed in order to maximise the board's dimensions. The bending stiffness degreases with increasing number of single timber members. The worst ratio between $E_{ef,par}$ and $E_{net,par}$ for long beams was determined for CLT beams with a ratio between h_r and h_p of 0.5 and with six parallel to the beam axis orientated boards. In this case, the ratio between $E_{ef,par}$ and $E_{net,par}$ is only 85%. Considering a beam thickness ratio of 75%, the bending stiffness for such a CLT beam is about 64% of the corresponding bending stiffness for a geometrical identical solid beam. Taking into account these results, long beams should be detailed either as DLT beams with side by side orientated diagonal layers or as CLT and DLT beams in general with a small number of timber members.

Fig. 126 to Fig. 131 contains the results for the bending edge stresses along path 1. The definition of path 1 is given in Fig. 31 and Fig. 32. All results are given as a ratio between σ_m and $\sigma_{m,net,par}$ in dependence of the number of parallel to the beam axis orientated boards. The bending edge stress σ_m was determined from the finite element analysis as the maximum value along path 1. Against it, $\sigma_{m,net,par}$ was calculated according to the beam theory taking into account the total thickness of the parallel layers and considering the parallel orientated layers as a full plane. Therefore following equation was used:

$$\sigma_{m,net,par} = \frac{M}{W_p} = \frac{6 \cdot F \cdot a}{\left(\sum h_p\right)^2 \cdot \sum t_p} \tag{18}$$

As long as the ratio between σ_m and $\sigma_{m,net,par}$ is exactly 100%, the beam can be considered as a geometrical identical solid beam with reduced beam thickness. Thereby, the beam thickness is equal to the total thickness of the parallel layers. With increasing ratio between σ_m and $\sigma_{m,net,par}$, the performance of CLT and DLT beams in terms of bending edge stresses degreases. In this case, the bending edge stresses are even higher than the bending edge stresses calculated according to the beam theory. Against it, the performance of the beams increases for degreasing ratio between σ_m and $\sigma_{m,net,par}$. In the special case, where the ratio between σ_m and $\sigma_{m,net,par}$ is equal to the ratio between the total thickness of the parallel layers and the total beam thickness ($\sigma_m / \sigma_{m,net,par} = t_p / t$), the performance of the CLT or DLT beam becomes equal to the performance of a geometrical identical solid beam.

To compare the different beam types together, the results for $t_p/t = 75\%$ (78%) were summarised in two diagrams. Fig. 43 contains the results for short beams, while in Fig. 44 results for long beams are given. Detailed results considering three different cases of t_p/t are given in the attachment in Fig. 126 to Fig. 131.

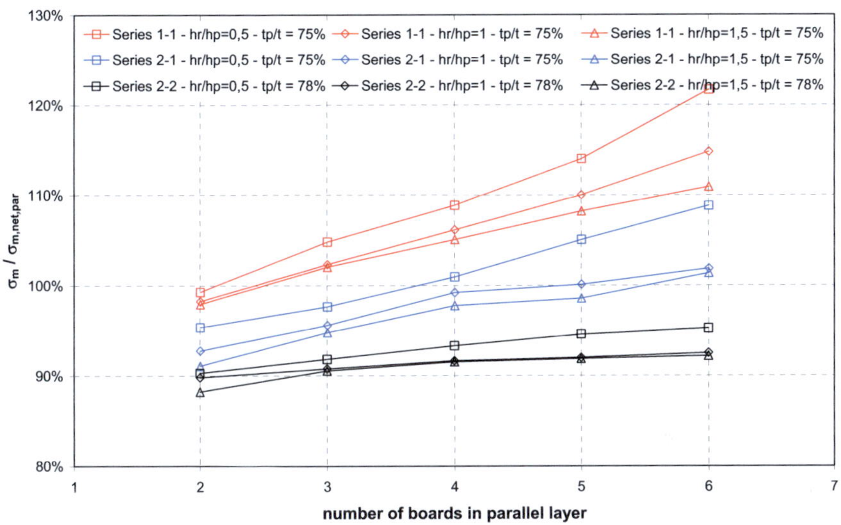

Fig. 43   Ratio between σ_m and $\sigma_{m,net}$ for short beams with $t_p/t = 75\%$ (78%)

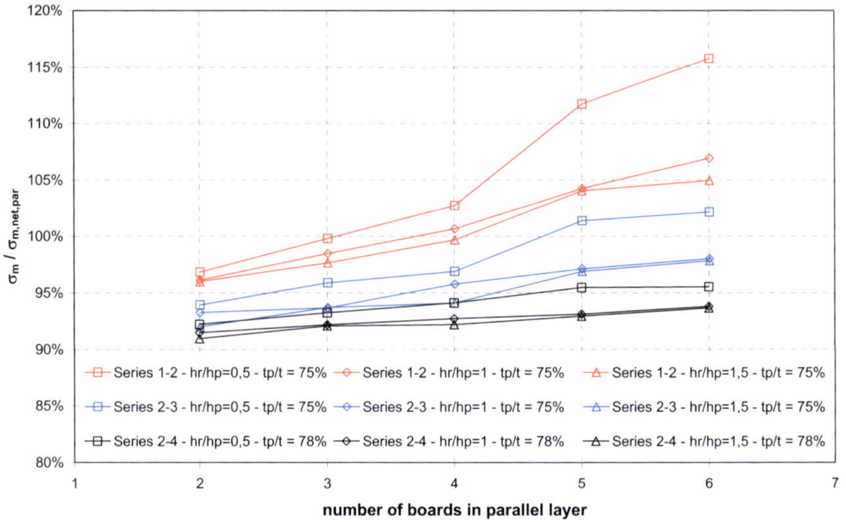

Fig. 44   Ratio between σ_m and $\sigma_{m,net}$ for long beams with $t_p/t = 75\%$ (78%)

For both, short and long beams, the performance of CLT beams is the worst while the performance of DLT beams with four plies is the best. Red lines represent the results for CLT beams. For short CLT beams, the ratio between σ_m and $\sigma_{m,net,par}$ was determined between 98% for $n_p = 2$ and 122% for $n_p = 6$. Against it, the ratio between σ_m and $\sigma_{m,net,par}$ for short DLT beams with four plies was determined between 88% for $n_p = 2$ and 95% for $n_p = 6$. The results for DLT beams with four plies are represented by black lines. All beam types behave similar: With increasing number of parallel to the beam axis orientated boards, the ratio between σ_m and $\sigma_{m,net,par}$ increases. Hence, the beam performance in terms of bending stresses degreases. However, the width h_r of the perpendicular or diagonal to the beam axis orientated boards influence the bending edge stresses, too. With degreasing width h_r, the ratio between σ_m and $\sigma_{m,net,par}$ increases.

Results for long beams (Fig. 44) are quite similar to those for short beams. For long CLT beams, the ratio between σ_m and $\sigma_{m,net,par}$ was determined between 96% for $n_p = 2$ and 116% for $n_p = 6$. Against it, the ratio between σ_m and $\sigma_{m,net,par}$ for long DLT beams with four plies was determined between 91% for $n_p = 2$ and 96% for $n_p = 6$. Hence, the results for long DLT beams with four plies are 5% to 20% smaller than the corresponding results for long CLT beams. Likewise for short beams, for long CLT and DLT beams the ratio between σ_m and $\sigma_{m,net,par}$ increases with increasing number n_p of parallel to the beam axis orientated boards.

To minimise bending stresses, beams loaded in bending should be produced using DLT instead of CLT elements. Those DLT elements should be made using side by side and in the opposite direction orientated boards. Furthermore, it is recommended to use DLT beams with a low number of boards with a large height. DLT beams with many narrow boards are not recommended.

Finally, shear stresses along path 2 were investigated using the finite element analysis. The definition of path 2 is given in Fig. 31 and Fig. 32. Path 2 contains the maximum shear stresses in the parallel to the beam axis orientated boards. All 270 results are given in diagrams in Fig. 132 to Fig. 137. In each diagram the results for each beam type with different t_p / t – ratios are given. To discuss the results, the following Fig. 45 and Fig. 46 contain only results for beams with a thickness ratio of t_p / t = 75% (78%). Fig. 45 contains results for short beams while results for long beams are given in Fig. 46.

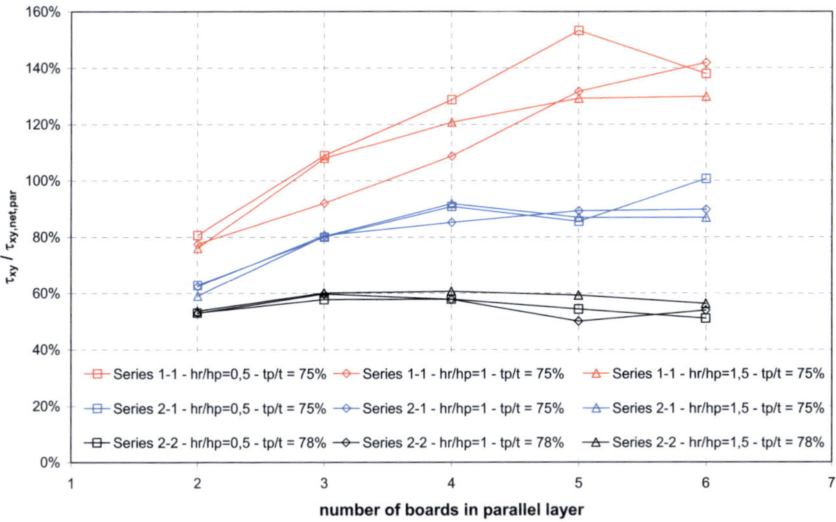

Fig. 45   Ratio between τ_{xy} and $\tau_{xy,net}$ for short beams with t_p/t = 75% (78%)

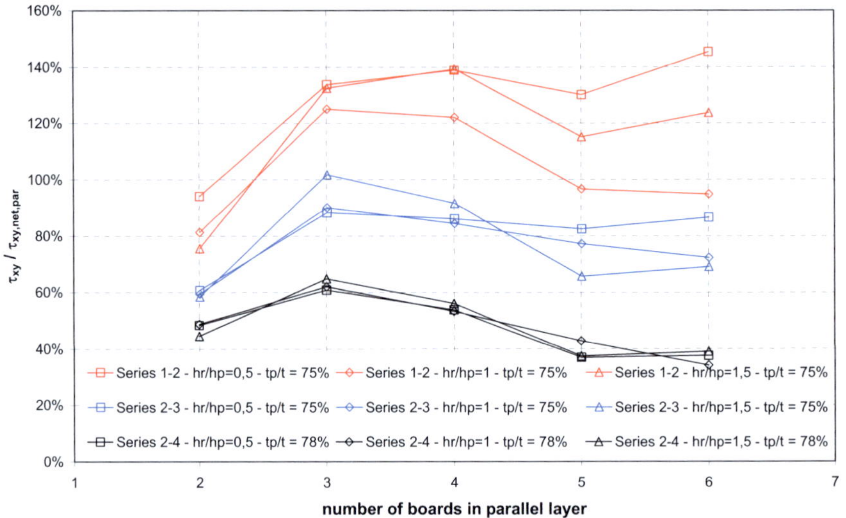

Fig. 46   Ratio between τ_{xy} and $\tau_{xy,net}$ for long beams with t_p/t = 75% (78%)

As in the previous diagrams, the red lines represent CLT beams (series 1-1 and 1-2), blue lines represent DLT beams with five plies (series 2-1 and 2-3) and results for DLT beams with four plies (series 2-2 and 2-4) are represented by black lines. Thereby, the ratio between the maximum shear stress τ_{xy} along path 2 taken from the finite element analysis and the maximum shear stress $\tau_{xy,net,par}$ calculated according to equation (19), in dependence on the number of parallel boards, is given.

$$\tau_{xy,net,par} = 1,5 \cdot \frac{F}{A_p} = \frac{F}{\sum h_p \cdot \sum t_p} \qquad (19)$$

The maximum shear stress $\tau_{xy,net,par}$ according to equation (19) corresponds to the shear stress for an equivalent solid beam with a beam thickness which equals to the thickness of the parallel layers in the CLT or DLT beams. The smaller the ratio between the shear stress τ_{xy} along path 2 and the maximum shear stress $\tau_{xy,net,par}$, the better is the performance of a beam. As expected, CLT beams are prone to fail in shear earlier than geometrical identical DLT beams with five or four plies. While for short CLT beams the ratio $\tau_{xy}/\tau_{xy,net,par}$ is between 76% and 153%, the ratio $\tau_{xy}/\tau_{xy,net,par}$ for short DLT beams with four layers is only between 54% and 61%. Similar situation can be observed for long beams. For long CLT beams the ratio $\tau_{xy}/\tau_{xy,net,par}$ is between 76% and 145%, while for long DLT beams with four layers the ratio $\tau_{xy}/\tau_{xy,net,par}$ is between 34% and 65%.

In the special case, where the ratio between τ_{xy} and $\tau_{xy,net,par}$ is equal to the ratio between the total thickness of the parallel layers and the total beam thickness ($\tau_{xy}/\tau_{xy,net,par}$ = t_p / t), the performance of the CLT or DLT beam becomes equal to the performance of a geometrical identical solid beam. In case, where the ratio between τ_{xy} and $\tau_{xy,net,par}$ is even smaller than the ratio t_p / t, the beam shear performance is even better than for a similar solid beam. This case occurs in particular for the investigated DLT beams with four plies.

As already mentioned, CLT and DLT beams are suitable for beams loaded in bending and by tensile stresses perpendicular to the beam axis direction. They are prone to transfer even high (tensile) loads perpendicular to the beam axis direction. Against it, comparable solid beams are prone to splitting. However, the bending performance of CLT and DLT beams is suboptimal compared to beams made of solid wood. Tests were made to investigate the load-carrying behaviour of CLT and DLT beams. Unfortunately due to the limited test specimen number, not all investigations were made and not all of the assumptions were proved. Hence, different finite element calculations were done. Using the finite element solid model, which fits the real CLT and DLT load-carrying behaviour the best, a parameter study was done. The finite element analysis shows, that the performance of CLT and DLT beams depends on the beam type and on the beam geometry. The best performance can be obtained using DLT beams. Thereby, the DLT beams should be produced with side by side and in the opposite direction orientated diagonal layers. Diagonal layers shouldn't be separated by parallel layers. The parallel to the beam axis orientated layers should be placed outside on the surface of the diagonal layers. Furthermore, the number of single timber members should be reduced to a minimum. With increasing number of single timber members, the stiffness and strength performance degreases. The necessary thickness of the single timber members and hence the ratio between t_r and t_p should be estimated in order to the ratio between the perpendicular to the beam axis orientated loads and the loads in the beam axis direction.

Using the diagrams in Fig. 114 to Fig. 137, the real stresses and the MOE values can be estimated for CLT and DLT beams. Using these diagrams, factors can be generated. These factors can be used to calculate maximum stresses and the MOE values for CLT and DLT beams loaded in bending based on the beam theory.

5 Summary and future prospects

The aim of this research work was to investigate the performance of cross laminated timber elements for beams loaded in bending and by tensile stresses perpendicular to the grain direction. Usually, beams loaded in bending are made out of solid wood or glulam. Both, solid wood and glulam, provide high strength and stiffness properties in grain direction while the strength and stiffness properties perpendicular to the grain direction are very small. For this reason, solid wood and glulam are less effective in beams, which are loaded both, in bending and by tensile stresses perpendicular to the grain direction. Examples for beams loaded perpendicular to the grain direction in tension are curved beams, beams with holes and beams with notched beam supports.

Cross laminated timber is a multilayer plate element with crosswise orientated layers. Usually, cross laminated timber is loaded as a plate in the out-of-plane direction or as a panel in the in-plane direction. Therefore, cross laminated timber is used for floors, for roofs and for wall systems. Due to the crosswise orientation of the layers, edgewise orientated CLT could be used as well for beams loaded both, in bending and by tensile stresses perpendicular to the grain direction. Thereby, parallel layers would transfer normal loads while the perpendicular to the beam axis orientated layers would act as reinforcement to transfer tensile stresses perpendicular to the beam axis direction.

Compared to geometrical identical solid wood or glulam beams, CLT beams are less stiff in bending. Hence, CLT as material for long beams is inefficient. The performance of long beams is usually governed by the bending stiffness. For this reason, a new and more effective element for beams loaded in bending was developed. Based on the idea of a CLT element with crosswise orientated boards, a diagonal orientated element (DLT) was investigated. Compared to CLT with crosswise orientated layers, the DLT element is a multilayer element with diagonal orientated layers. At least two diagonal layers are necessary to make a symmetric DLT beam. The diagonal layers must be orientated in couples and in the opposite direction to each other. Within this research work, two cases for DLT beams were investigated: DLT elements with diagonal layers, which are separated by a parallel layer and DLT elements with side by side orientated diagonal layers were investigated. Likewise CLT elements, DLT elements contain parallel orientated layers, too.

Within this research work, beams were manufactured using self-made CLT and DLT elements. To enforce shear failure, short CLT and DLT beams were made. Long CLT and DLT were made to enforce bending failure. For beams made using DLT elements, two cases were studied: Beams were made either with DLT elements with separated diagonal layers or with DLT elements with side by side orientated diagonal layers. Altogether 60 short and long, CLT and DLT beams with four and five plies were produced.

All beams were tested in bending. 30 tests were done on unchanged beams while 30 tests were done on beams with notched beam supports or with holes. The results for beams with notched beam supports and with holes are nearly similar to those for beams without any singularities. Using CLT and DLT instead of solid wood, brittle timber failure could be prevented. The beams with notched beam supports and with holes were designed to enforce timber splitting at lower loads. Due to the perpendicular or diagonal reinforcing layers, timber splitting was prevented very well.

Unfortunately, the test results couldn't be used to qualify, whether CLT or DLT elements are better for beams loaded in bending. To quantify CLT and DLT beams, a finite element analysis was performed. First, different numerical models were studied. A beam, a plane and a solid model were discussed. Governing parameters were investigated and those, which don't affect the load-displacement behaviour, were neglected. Finally, the solid model was used for a parameter study. The solid model fits the load-displacement behaviour of CLT and DLT beams the best. Numerical calculations were done on totally 270 beams. Thereby, it was clearly demonstrated, that DLT elements provides a higher performance than CLT elements for beams loaded in bending. However, this result is more valid for DLT beams with side by side and in the opposite direction and diagonal to the beam axis orientated layers. Against it, diagonal orientated layers, which are separated by at least one parallel layer, are less good.

This work doesn't provide general equations to calculate the strength and stiffness properties for CLT and DLT beams loaded in bending. This work rather shows how to calculate CLT and DLT beams using a finite element analysis and which model fits the real load-displacement behaviour the best. Finally, this work presents a completely new, until now not investigated, DLT product, which can be used very well for beams loaded both, in bending and by tensile stresses perpendicular to the grain direction.

6 References

[1] DIN 1052: 2004-08. Entwurf, Berechnung und Bemessung von Holzbauwerken – Allgemeine Bemessungsregeln und Bemessungsregeln für den Hochbau. DIN Deutsches Institut für Normung e.V., Beuth Verlag GmbH, 10772 Berlin

[2] Blaß, H.J., Görlacher, R. (2002). „Zum Trag- und Verformungsverhalten von LIGNOTREND Elementen bei Beanspruchung in Plattenebene", Bauen mit Holz 104 (2002); no. 11, page 34-41; no 12, page 30-34

[3] Möhler, K. (1956): Über das Tragverhalten von Biegeträgern und Druckstäben mit zusammengesetztem Querschnitt und nachgiebigen Verbindungsmitteln. Habilitation. Karlsruhe, Juni 1956

[4] Blaß, H.J., Görlacher, R. (2001). „Zum Trag- und Verformungsverhalten von LIGNOTREND-Decken und Wandsystemen aus Nadelschnittholz", Bauen mit Holz 103 (2001); no. 4, page 37-40; no 5, page 68-71

[5] Görlacher, R. (2002). „Ein Verfahren zur Bestimmung des Rollschubmoduls von Holz", Holz als Roh- und Werkstoff 60 (2002); Springer Verlag, page 317-322

[6] Kreuzinger, H. (1999). „Platten, Scheiben und Schalen – ein Berechnungsmodell für gängige Statikprogramme", Bauen mit Holz (1999); no. 1, page 34-39

[7] Kreuzinger, H. (2000). „Verbundkonstruktionen aus nachgiebig miteinander verbundenen Querschnittsteilen", in Tagungsband 2000 Ingenieurholzbau – Karlsruher Tage, Bruderverlag, Karlsruhe

[8] Brettsperrholz – Ein Blick auf Forschung und Entwicklung. Tagungsband zum 5. Grazer Holzbau Fachtagung 2006. GraHFT'06. Herausgeber: Institut für Holzbau und Holztechnologie, holz.bau forschungs GmbH, Institut für Stahlbau und Flächentragwerke, Technische Universität Graz, Austria

[9] Uibel, T., Blaß, H.J. (2006). „Load carrying capacity of joints with dowel type fasteners in solid wood panels". International council for research and innovation in building and construction. Work commission W18 – Timber structures. CIB-W18, paper 39-7-5, Meeting 39, Florence, Italy 2006

[10] Tobisch, S. (2006). "Methoden zur Beeinflussung ausgewählter Eigenschaften von dreilagigen Massivholzplatten aus Nadelholz". Dissertation der Universität Hamburg, Fachbereich Biologie

[11] Bosl, R. (2002). „Zum Nachweis des Trag- und Verformungsverhaltens von Wandscheiben aus Brettlagenholz". Dissertation der Universität der Bundeswehr München, Fakultät für Bauingenieur- und Vermessungswesen, Konstruktive Gestaltung und Holzbau

[12] DIN EN 789. Holzbauwerke – Prüfverfahren – Bestimmung der mechanischen Eigenschaften von Holzwerkstoffen; Deutsche Fassung DIN EN 789:2004. DIN Deutsches Institut für Normung e.V., Beuth Verlag GmbH, 10772 Berlin

[13] Blaß, H.J., Bejtka, I. (2003). "Querzugverstärkungen in gefährdeten Bereichen mit selbstbohrenden Holzschrauben" Forschungsbericht 2003. Versuchsanstalt für Stahl, Holz und Steine, Universität Karlsruhe

7 Appendix

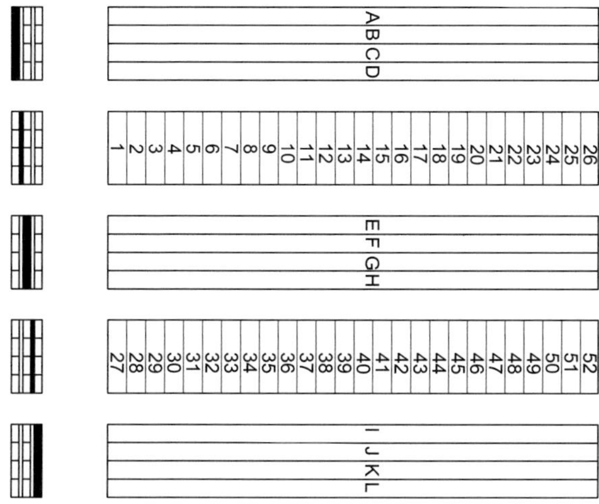

Fig. 47   Species 1.1 – Short beam with 90° reinforcing layers - 5 ply

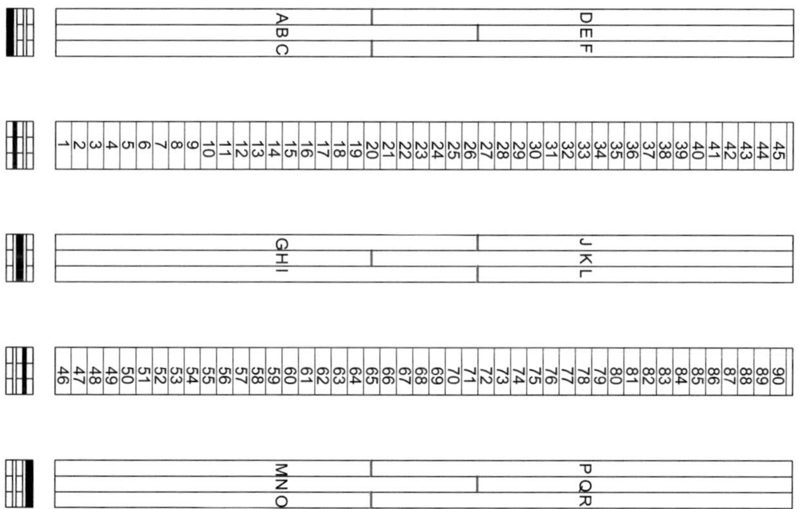

Fig. 48   Species 1.2 – Long beam with 90° reinforcing layers - 5 ply

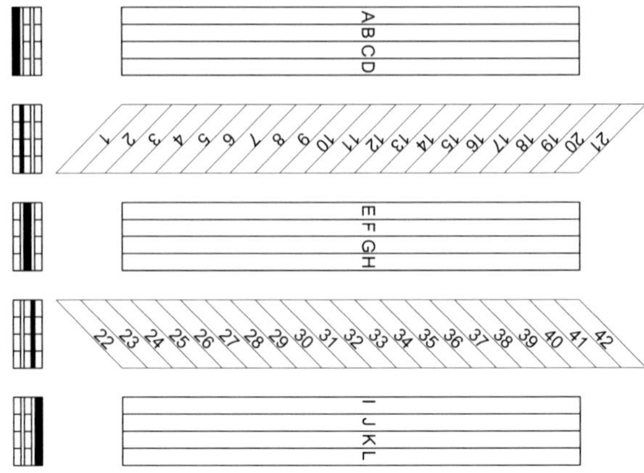

Fig. 49   Species 2.1 – Short beam with 45° reinforcing layers - 5 ply

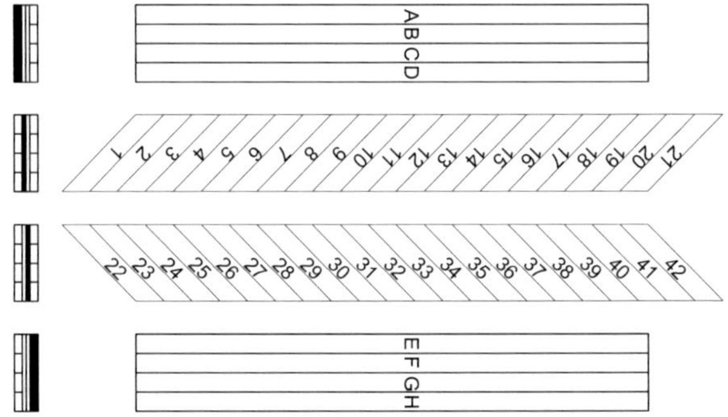

Fig. 50   Species 2.2 – Short beam with 45° reinforcing layers - 4 ply

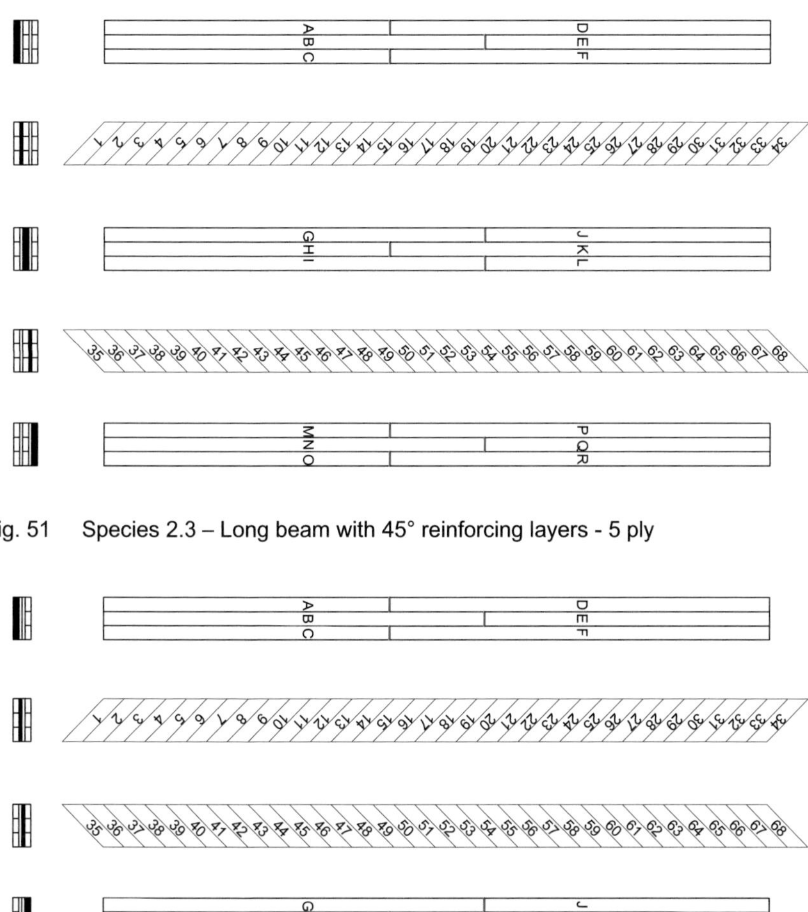

Fig. 51   Species 2.3 – Long beam with 45° reinforcing layers - 5 ply

Fig. 52   Species 2.4 – Long beam with 45° reinforcing layers - 4 ply

Fig. 53   Species 1.1 – Application of the PRF glue on the first parallel layer

Fig. 54   Species 1.1 – Left: Assembling of the first reinforcing layer (90°). Right: Assembling of the second parallel layer

Fig. 55 Species 1.1 – Application of the PRF glue on the second parallel layer

Fig. 56 Species 1.1 – Beam with a fixing frame on the way to the press

Fig. 57   Species 1.1 – Two beams in a fixing frame on the way to the press

Fig. 58   Species 1.2 – Beams after pressing

Fig. 59 Species 2.5 – Diagonal reinforcing layer with a PRF glue surface

Fig. 60 Species 2.4 – Assembling of the diagonal layer on the first parallel layer

Fig. 61   Species 2.4 – Assembling of the second diagonal layer

Fig. 62   Species 2.4 – Assembling of the second diagonal layer

Fig. 63   Species 2.4 – Assembling of the second parallel layer

Fig. 64   Species 2.4 – Beam in a fixing frame short before on the way to the press

Fig. 65   Species 2.3 – Beams after pressing

Fig. 66   Species 1.2 and 2.3 – Beams after pressing

Table 14   Allocation of parameters in species 1-1, numeration displayed in Fig. 47

board \ beam	Dynamic MOE in kN/mm² - Dry timber density in kg/m³											
	1	2	3	4	5	6	7	8	9	10	11	12
1	11 - 526	11,2 - 376	12,3 - 539	6,26 - 471	9,65 - 445	9,07 - 390	9,99 - 474	11,1 - 452	11,3 - 429	7,51 - 410	8,61 - 488	8,61 - 488
2	10,2 - 468	10,6 - 432	11,8 - 407	10,2 - 463	9,65 - 445	9,06 - 463	10,9 - 463	9,69 - 446	9,69 - 446	8,1 - 479	10,3 - 444	10,3 - 444
3	11,1 - 470	6,73 - 386	7,72 - 411	7,39 - 461	10,8 - 467	12,2 - 453	12,3 - 456	9,97 - 421	9,97 - 421	6,24 - 477	9,79 - 382	6,41 - 408
4	8,94 - 439	11,3 - 481	10,9 - 446	9,73 - 446	8,55 - 456	11,3 - 487	10,9 - 463	9,6 - 467	10,7 - 411	7,25 - 455	12,1 - 467	15,1 - 534
5	11,9 - 455	10,7 - 442	6,26 - 471	8,01 - 485	10,4 - 473	8,78 - 422	9,71 - 451	10,7 - 488	10,7 - 411	6,24 - 477	12,1 - 467	15,1 - 534
6	9,49 - 418	11,2 - 376	9,22 - 420	10,6 - 431	9,65 - 548	8,29 - 432	9,69 - 446	10,7 - 411	11,1 - 452	10,7 - 472	9,79 - 382	8,61 - 488
7	8,76 - 442	11,8 - 486	9,18 - 481	7,5 - 448	9,65 - 445	6,94 - 401	8,7 - 418	7,52 - 436	10,7 - 488	7,52 - 436	10,7 - 454	14,6 - 524
8	11,2 - 507	11,2 - 376	6,47 - 417	7,72 - 411	12,9 - 478	11,2 - 376	7,75 - 428	9,61 - 453	13,4 - 484	11,3 - 456	6,41 - 408	9,79 - 382
9	10,4 - 473	6,98 - 480	10,6 - 481	11,8 - 407	11 - 526	9,07 - 390	7,52 - 436	6,24 - 477	7,41 - 437	7,03 - 395	12,1 - 467	10,8 - 450
10	8,55 - 456	8,29 - 432	14,8 - 485	14,8 - 485	8,94 - 439	10,6 - 432	9,18 - 481	7,75 - 428	12,3 - 456	14,5 - 488	9,79 - 382	9,83 - 447
11	12,9 - 478	6,94 - 401	14,4 - 487	10,2 - 463	8,76 - 442	7,1 - 441	8,84 - 561	10,2 - 464	10,9 - 463	9,6 - 467	8,96 - 445	14,6 - 524
12	9,65 - 548	9,67 - 472	10,9 - 463	9,61 - 449	8,16 - 423	10,1 - 462	5,9 - 448	9,6 - 467	12,3 - 456	5,9 - 448	12,8 - 555	6,41 - 408
13	9,49 - 418	10,7 - 442	11,8 - 407	7,5 - 448	8,55 - 456	13,2 - 481	11,3 - 429	7,75 - 428	7,41 - 437	11,6 - 499	8,14 - 418	14,6 - 524
14	8,76 - 442	9,06 - 463	7,5 - 448	6,26 - 471	10,8 - 467	9,06 - 463	11,1 - 452	7,51 - 410	13,4 - 484	5,9 - 448	9,79 - 382	8,61 - 488
15	8,94 - 439	8,78 - 422	12,3 - 539	12,3 - 539	11,1 - 470	10,7 - 442	12,6 - 461	10,2 - 425	11,3 - 429	10,2 - 425	10,8 - 450	8,14 - 418
16	8,36 - 430	13,2 - 481	11,2 - 512	6,98 - 480	8,93 - 406	9,06 - 463	5,9 - 448	9,11 - 425	9,96 - 524	11,3 - 456	12,8 - 555	10,3 - 489
17	8,93 - 406	9,06 - 463	10,9 - 446	9,61 - 449	8,36 - 430	11,3 - 481	7,75 - 428	9,69 - 446	9,6 - 467	13,4 - 484	8,96 - 445	8,99 - 416
18	11,1 - 470	8,78 - 422	6,47 - 417	9,22 - 420	8,97 - 423	6,94 - 401	7,25 - 455	9,97 - 421	8,7 - 418	10,7 - 488	9,83 - 447	7,82 - 418
19	8,97 - 423	12,2 - 453	6,47 - 417	12,2 - 453	11 - 526	10,7 - 442	9,28 - 395	6,24 - 477	8,16 - 396	9,97 - 421	7,66 - 380	10,3 - 489
20	8,16 - 423	8,29 - 432	10,2 - 463	7,5 - 448	14,1 - 540	9,07 - 390	10,3 - 477	10,2 - 425	10,9 - 463	9,6 - 467	8,96 - 445	9,83 - 447
21	10,3 - 442	6,98 - 480	9,22 - 420	10,2 - 463	10,4 - 473	6,73 - 386	11,3 - 429	14,5 - 488	9,97 - 421	6,24 - 477	10,3 - 489	6,41 - 408
22	14,1 - 540	7,1 - 441	8,01 - 485	10,6 - 431	8,55 - 456	10,6 - 432	14,5 - 488	7,41 - 437	7,71 - 415	10,2 - 425	8,99 - 416	8,14 - 418
23	9,65 - 548	10,1 - 462	9,99 - 474	9,22 - 420	9,65 - 445	11,3 - 481	8,16 - 396	11,3 - 456	7,51 - 410	11,1 - 452	8,96 - 445	8,99 - 416
24	10,8 - 467	7,1 - 446	9,73 - 446	10,9 - 446	9,65 - 548	12,2 - 453	7,71 - 415	9,61 - 453	9,96 - 524	13,4 - 484	6,41 - 408	8,99 - 416
25	10,2 - 468	9,67 - 472	9,61 - 449	11,2 - 512	10,3 - 442	11,2 - 376	11,6 - 499	11,3 - 456	10,7 - 472	9,11 - 425	8,64 - 409	8,14 - 418
26	11,9 - 455	14,4 - 487	9,73 - 446	6,26 - 471	8,97 - 423	10,2 - 468	10,3 - 477	14,5 - 488	10,3 - 477	9,71 - 451	7,82 - 418	10,3 - 489
27	11,2 - 507	11,2 - 376	7,25 - 455	9,99 - 474	8,36 - 430	11,2 - 376	9,98 - 472	10,7 - 411	9,98 - 472	10,7 - 488	14,6 - 524	8,61 - 488
28	8,97 - 423	8,01 - 485	9,96 - 524	8,78 - 422	10,3 - 442	11,8 - 486	11,6 - 499	11,1 - 452	11,3 - 567	7,51 - 410	10,7 - 454	10,7 - 454
29	10,3 - 442	10,6 - 481	7,03 - 395	8,78 - 422	10,2 - 468	10,6 - 546	8,7 - 418	10,2 - 464	7,39 - 461	14,5 - 488	8,64 - 409	10,7 - 454
30	9,65 - 445	6,26 - 471	9,18 - 481	9,06 - 463	8,93 - 406	6,98 - 480	10,7 - 472	9,11 - 425	9,96 - 524	10,3 - 477	12,8 - 555	12,1 - 467
31	8,55 - 456	7,5 - 448	6,47 - 417	10,1 - 462	9,49 - 418	7,1 - 441	11,6 - 499	9,97 - 421	7,71 - 415	11,3 - 456	8,61 - 488	12,1 - 467
32	12,9 - 478	9,61 - 449	10,6 - 481	7,1 - 446	11,1 - 470	10,1 - 462	9,71 - 451	12,3 - 456	12,6 - 461	9,6 - 467	8,14 - 418	10,3 - 444
33	9,65 - 548	10,9 - 463	9,22 - 420	13,2 - 481	11,9 - 455	7,1 - 446	10,7 - 472	13,4 - 484	8,28 - 445	9,61 - 453	10,7 - 454	10,7 - 454
34	10,4 - 473	10,6 - 481	14,8 - 485	8,29 - 432	9,49 - 418	9,67 - 472	9,71 - 451	12,3 - 456	10,9 - 463	8,7 - 418	8,64 - 409	15,1 - 534
35	11,1 - 470	12,3 - 539	10,3 - 477	10,9 - 446	8,76 - 442	10,6 - 546	10,9 - 463	9,11 - 425	8,84 - 561	7,25 - 455	12,8 - 555	15,1 - 534
36	11,9 - 455	11,3 - 487	11,2 - 512	7,72 - 411	10,8 - 467	7,1 - 441	13,4 - 484	8,1 - 479	9,98 - 472	8,84 - 561	8,14 - 418	9,79 - 382
37	10,8 - 407	14,4 - 407	11,3 - 567	11,2 - 512	11,2 - 507	11,8 - 486	10,0 - 463	8,1 - 479	9,98 - 472	7,71 - 415	10,8 - 450	10,8 - 450
38	8,76 - 442	9,67 - 472	7,72 - 411	11,2 - 512	10,2 - 468	11,2 - 376	10,7 - 488	11,1 - 452	7,39 - 461	10,7 - 488	8,64 - 409	9,64 - 439
39	9,49 - 418	10,6 - 431	9,18 - 481	7,72 - 411	8,16 - 423	10,6 - 432	9,98 - 472	7,75 - 428	8,28 - 445	10,2 - 425	9,83 - 447	7,66 - 380
40	8,93 - 406	10,2 - 463	8,01 - 485	11,8 - 407	8,76 - 442	11,3 - 481	8,28 - 445	7,51 - 410	8,16 - 396	12,6 - 461	7,66 - 380	15,1 - 534
41	8,36 - 430	10,6 - 431	9,73 - 446	11,8 - 407	8,55 - 456	8,78 - 422	8,84 - 561	7,03 - 395	8,16 - 396	9,69 - 446	7,82 - 418	10,3 - 444
42	10,3 - 442	14,8 - 485	10,6 - 431	10,6 - 481	9,65 - 548	11,3 - 487	8,7 - 418	10,7 - 411	7,39 - 461	8,1 - 479	8,64 - 409	10,8 - 450
43	8,97 - 423	11,8 - 486	10,2 - 463	8,29 - 432	12,9 - 478	11,2 - 376	8,84 - 561	10,2 - 464	7,52 - 436	10,7 - 411	9,83 - 447	12,8 - 555
44	11 - 526	10,6 - 546	9,61 - 449	6,47 - 417	10,4 - 473	13,2 - 481	8,16 - 396	10,2 - 464	11,6 - 499	12,3 - 456	7,66 - 380	8,99 - 416
45	10,1 - 462	6,98 - 480	14,8 - 485	10,9 - 463	10,3 - 442	9,67 - 472	8,7 - 418	7,03 - 395	10,7 - 472	9,71 - 451	8,96 - 445	7,66 - 380
46	11,3 - 481	6,73 - 386	12,3 - 539	9,99 - 474	11 - 526	10,7 - 442	9,28 - 395	7,71 - 415	9,71 - 451	9,96 - 524	10,8 - 450	14,6 - 524
47	10,6 - 432	10,6 - 432	11,2 - 512	14,8 - 485	14,1 - 540	11,2 - 376	11,3 - 429	6,24 - 477	7,25 - 455	7,75 - 428	8,64 - 409	12,8 - 555
48	13,2 - 481	10,6 - 481	10,9 - 446	12,3 - 539	10,2 - 468	11,8 - 486	12,6 - 461	11,6 - 499	9,61 - 453	9,61 - 453	10,3 - 444	9,83 - 447
49	7,1 - 441	9,73 - 446	7,72 - 411	6,47 - 417	11,1 - 470	12,2 - 453	7,25 - 455	10,2 - 464	9,28 - 395	7,71 - 415	8,96 - 445	8,99 - 416
50	10,6 - 546	11,2 - 376	11,8 - 407	10,9 - 463	8,93 - 406	6,98 - 480	5,9 - 448	9,69 - 446	10,3 - 477	10,7 - 472	10,3 - 489	14,6 - 524
51	11,8 - 486	9,07 - 390	10,9 - 463	7,1 - 441	8,97 - 423	10,1 - 462	5,9 - 448	10,2 - 425	7,41 - 437	14,5 - 488	10,3 - 489	7,82 - 418
52	12,9 - 478	9,98 - 472	9,28 - 395	8,01 - 485	9,49 - 418	8,16 - 396	7,39 - 461	7,51 - 410	9,28 - 395	9,61 - 453	10,3 - 444	7,66 - 380

A	13,2 - 438	13,8 - 453	13,6 - 480	11,3 - 449	11,6 - 467	11,1 - 420	12,2 - 466	11,9 - 461	14,2 - 472	10,9 - 490	10,1 - 430	9,75 - 436
B	10,4 - 427	10,3 - 421	12,1 - 442	12,1 - 441	11,4 - 449	10,4 - 458	10,9 - 461	11,6 - 482	11,4 - 437	10,9 - 424	14,4 - 435	12,7 - 451
C	10,2 - 440	12,1 - 454	10,6 - 421	13,5 - 505	12,3 - 475	12,6 - 503	10,4 - 402	14,1 - 479	11,6 - 441	13,6 - 460	13 - 455	11 - 359
D	11,4 - 433	12,1 - 414	13,7 - 462	13,1 - 440	13,2 - 464	14,5 - 520	13,3 - 431	10,2 - 418	13,6 - 466	10,1 - 434	11,4 - 389	15 - 478
E	14,4 - 485	12,2 - 468	13,4 - 502	10,5 - 454	9,41 - 430	13,3 - 470	13,1 - 434	12,2 - 447	11 - 438	11,8 - 486	14,4 - 440	10,7 - 375
F	11,2 - 451	9,45 - 416	13,6 - 477	10,8 - 404	12,9 - 461	12,1 - 438	10,9 - 426	10,5 - 433	12,1 - 421	8,48 - 428	10,9 - 484	15,1 - 434
G	14,8 - 518	11,1 - 479	10,7 - 437	11,5 - 438	11,5 - 457	11,4 - 431	10,1 - 422	9,22 - 407	12,5 - 481	11,9 - 452	9,35 - 424	12,4 - 452
H	14,3 - 518	10,1 - 413	13,3 - 431	10,6 - 421	11,4 - 440	12,9 - 453	10,9 - 426	10,2 - 431	15,6 - 540	12,4 - 464	14,4 - 462	12,4 - 469
I	12,1 - 454	12,1 - 456	12,2 - 452	10,4 - 424	14,2 - 472	9,99 - 409	10,8 - 440	11,6 - 449	11,5 - 425	11,7 - 450	10,1 - 429	10,8 - 367
J	11,1 - 427	12 - 452	9,24 - 435	9,77 - 440	11,4 - 419	13,5 - 473	12,1 - 450	9,98 - 432	13 - 503	12,9 - 529	10,4 - 450	9,67 - 375
K	8,97 - 436	12,3 - 417	9,26 - 403	11,7 - 447	10,6 - 439	11,3 - 428	13,6 - 484	11,3 - 437	10,7 - 429	12,4 - 495	15,5 - 516	9,66 - 456
L	10,8 - 456	13,9 - 460	9,31 - 447	12,4 - 486	11,5 - 451	14,9 - 486	10,1 - 408	11,8 - 446	12,9 - 458	10,8 - 427	10,6 - 361	15,1 - 465

Table 15   Allocation of parameters in species 1-2, numeration displayed in Fig. 48

board \ beam	1	2	3	4	5	6	7	8	9	10	11	12
1	10,4 - 467	11,7 - 415	12,6 - 480	6,17 - 395	11,2 - 431	11,2 - 431	10,6 - 465	8,14 - 423	8,97 - 439	11,2 - 536	9,12 - 416	10,6 - 464
2	11,7 - 462	9,89 - 429	7,38 - 399	6,17 - 395	9,62 - 470	8,97 - 439	9,89 - 429	8,56 - 459	11,2 - 431	5,92 - 466	8,41 - 419	7,2 - 475
3	8,49 - 442	6,5 - 398	6,95 - 465	6,17 - 395	8,76 - 369	9,62 - 470	5,34 - 355	8,68 - 436	9,39 - 384	10,7 - 461	10,9 - 483	7,24 - 444
4	8,39 - 412	5,34 - 355	7,41 - 371	13,6 - 564	11,6 - 443	8,97 - 439	6,5 - 398	8,15 - 440	8,9 - 395	9,42 - 416	6,39 - 432	8,09 - 409
5	9,73 - 448	5,34 - 355	9,51 - 488	13,6 - 564	10,7 - 449	11,2 - 431	13,1 - 424	9,47 - 370	11,6 - 443	7,13 - 445	5,68 - 429	8,45 - 385
6	9,73 - 448	9,73 - 448	13,2 - 505	7,87 - 493	11,1 - 430	8,9 - 395	6,56 - 373	11,1 - 463	5,22 - 448	10,7 - 461	8,68 - 436	14,9 - 503
7	10,4 - 467	9,73 - 448	9,28 - 495	7,87 - 493	6,69 - 434	6,69 - 434	9,87 - 399	9,12 - 416	7,15 - 489	7,13 - 445	5,54 - 425	8,14 - 426
8	12,1 - 451	11,7 - 462	6,95 - 465	7,87 - 493	8,93 - 414	5,22 - 448	10,9 - 452	5,68 - 429	9,39 - 384	10,2 - 475	8,15 - 440	8,52 - 409
9	9,37 - 469	10,4 - 467	7,38 - 399	12,7 - 434	10,7 - 461	8,86 - 479	13,1 - 424	10,9 - 418	8,86 - 479	6,78 - 393	10,9 - 418	8,9 - 466
10	13,2 - 463	8,49 - 442	12,6 - 480	12,7 - 434	9,42 - 416	9,95 - 428	8,23 - 398	8,68 - 436	11,2 - 431	11,2 - 495	9,47 - 370	10,7 - 446
11	9,28 - 495	8,39 - 412	7,41 - 371	10,7 - 500	6,14 - 327	9,49 - 456	9,89 - 429	5,54 - 425	8,97 - 439	5,91 - 451	9,29 - 435	10,7 - 446
12	9,43 - 389	7,46 - 445	9,28 - 453	10,7 - 500	9,39 - 384	6,39 - 420	10,6 - 465	10,9 - 418	9,93 - 403	7,13 - 434	6,91 - 418	9,57 - 481
13	12,1 - 451	7,84 - 410	10,1 - 489	9,92 - 418	8,93 - 414	9,62 - 470	5,34 - 355	6,2 - 515	6,39 - 420	9,42 - 416	8,14 - 423	8,87 - 450
14	9,37 - 469	8,24 - 442	10,4 - 467	6,17 - 395	7,38 - 477	9,39 - 384	6,5 - 398	9,47 - 370	9,95 - 428	8,9 - 452	4,51 - 425	9,09 - 463
15	9,28 - 495	9,32 - 440	7,46 - 445	6,17 - 395	8,97 - 439	7,13 - 445	8,17 - 431	8,9 - 410	10,7 - 449	9 - 471	8,9 - 410	6,23 - 426
16	7,46 - 445	12,1 - 451	7,46 - 445	8,25 - 397	9,95 - 428	9,42 - 416	8,78 - 427	7,93 - 461	11,1 - 430	5,92 - 466	8,56 - 459	8,18 - 413
17	7,84 - 410	9,37 - 469	9,73 - 448	10,7 - 500	9,49 - 456	8,93 - 414	11 - 387	4,51 - 425	8,93 - 414	7,34 - 471	10,1 - 441	9,1 - 459
18	9,32 - 440	9,28 - 495	12,6 - 480	8,49 - 442	5,22 - 448	8,76 - 369	11,9 - 418	11,5 - 407	8,76 - 369	7,39 - 455	8,15 - 417	9,48 - 438
19	8,24 - 442	9,43 - 389	8,24 - 442	11,7 - 462	8,86 - 479	11,6 - 443	10 - 410	8,15 - 417	11,6 - 443	10,7 - 449	9,42 - 425	9,08 - 418
20	13,2 - 505	12,1 - 451	6,95 - 465	13,4 - 538	8,9 - 395	10,7 - 461	11,2 - 428	6,28 - 400	7,39 - 455	12,4 - 522	11,5 - 407	8,95 - 407
21	10,4 - 467	9,37 - 469	6,95 - 465	13,4 - 538	8,18 - 413	9,62 - 470	8,58 - 373	8,14 - 423	7,38 - 477	8,22 - 414	6,91 - 418	10,6 - 464
22	11,7 - 462	8,77 - 458	7,41 - 371	9,92 - 418	6,39 - 420	7,15 - 489	13,1 - 424	10,9 - 483	8,86 - 479	6,25 - 406	9,38 - 405	11 - 440
23	8,49 - 442	11,7 - 485	10,8 - 493	7,87 - 493	7,39 - 455	9,95 - 428	8,23 - 398	8,41 - 419	9,93 - 403	9,71 - 420	10,9 - 483	8,09 - 409
24	8,39 - 412	9,58 - 481	12,6 - 480	8,25 - 397	6,39 - 420	9,49 - 456	9,89 - 429	11,1 - 463	6,69 - 434	8,74 - 467	6,39 - 432	8,45 - 385
25	9,28 - 453	9,51 - 488	7,38 - 399	8,25 - 397	8,87 - 450	6,39 - 420	12,4 - 413	6,91 - 418	8,9 - 395	10,2 - 502	8,9 - 410	8,9 - 466
26	11,7 - 485	8,39 - 412	10,8 - 493	6,17 - 395	7,24 - 444	7,38 - 477	11,2 - 428	5,68 - 429	7,15 - 489	7,34 - 471	5,68 - 429	8,52 - 409
27	8,77 - 458	11,7 - 462	11,7 - 462	7,97 - 387	6,69 - 434	6,69 - 434	12,4 - 413	9,38 - 405	11,2 - 536	9,89 - 474	9,29 - 435	14,9 - 503
28	9,32 - 440	8,49 - 442	9,32 - 440	8,25 - 397	7,38 - 477	8,86 - 479	10,9 - 452	5,54 - 425	6,78 - 393	11,8 - 521	8,15 - 417	8,14 - 426
29	7,84 - 410	9,28 - 495	6 - 360	7,97 - 387	5,22 - 448	7,38 - 477	8,23 - 398	8,15 - 440	11,8 - 521	11,8 - 521	9,42 - 425	9,08 - 418
30	7,46 - 445	9,43 - 389	11,7 - 485	7,97 - 387	8,9 - 395	5,22 - 448	11,7 - 415	10,9 - 418	7,34 - 471	12,4 - 522	11,5 - 407	8,95 - 407
31	9,28 - 453	10,9 - 452	9,58 - 481	18,2 - 551	8,45 - 385	9,95 - 428	9,87 - 399	9,12 - 416	11,2 - 495	10,2 - 502	8,15 - 440	9,1 - 459
32	13,2 - 463	9,89 - 429	7,38 - 399	7,46 - 440	10,7 - 461	8,86 - 479	13,1 - 424	8,68 - 436	9 - 471	9,71 - 420	10,9 - 418	7,24 - 444
33	8,24 - 442	6,5 - 398	18,2 - 551	9,92 - 418	8,09 - 409	9,49 - 456	5,34 - 355	9,12 - 416	8,9 - 452	5,91 - 451	9,47 - 370	9,57 - 481
34	8,77 - 458	9,73 - 448	9,28 - 495	18,2 - 551	11 - 440	7,15 - 489	10,6 - 465	10,9 - 483	10,2 - 475	7,13 - 434	8,56 - 459	7,2 - 475
35	11,7 - 485	10,4 - 467	12,6 - 480	8,49 - 442	10,6 - 464	6,39 - 420	6,5 - 398	8,41 - 419	5,91 - 451	6,86 - 424	6,28 - 400	9,57 - 484
36	9,58 - 481	8,54 - 439	7,38 - 399	12,7 - 434	8,9 - 466	10,7 - 449	9,38 - 405	9,47 - 370	6,25 - 406	8,74 - 467	7,93 - 461	10,7 - 446
37	6,52 - 417	12,4 - 413	12,6 - 480	11 - 465	8,52 - 409	11,1 - 430	9,12 - 416	11,2 - 428	8,70 - 309	10,2 - 502	8,41 - 419	9,57 - 481
38	13,2 - 505	7,46 - 445	6 - 360	7,97 - 387	14,9 - 503	8,93 - 414	11,1 - 463	10 - 410	8,93 - 414	9 - 471	11,1 - 463	8,87 - 450
39	9,58 - 481	9,28 - 453	13,4 - 538	12,7 - 434	8,14 - 426	8,76 - 369	10,1 - 441	5,34 - 355	8,22 - 414	11,2 - 536	6,39 - 432	6,23 - 426
40	9,51 - 488	8,77 - 458	7,84 - 410	12,7 - 434	14,9 - 503	11,6 - 443	6,28 - 400	9,89 - 429	6,25 - 406	10,3 - 453	10,9 - 483	8,18 - 413
41	9,28 - 453	11,7 - 485	6 - 360	18,2 - 551	10,6 - 464	8,97 - 439	9,42 - 425	11 - 482	11,1 - 430	11,2 - 495	5,54 - 425	9,09 - 463
42	6,52 - 417	9,32 - 440	10,8 - 493	7,97 - 387	8,45 - 385	11,2 - 431	11,5 - 407	8,56 - 459	7,13 - 434	10,2 - 475	8,68 - 436	8,52 - 409
43	9,58 - 481	10,1 - 489	9,58 - 481	7,97 - 387	9,62 - 470	9,39 - 384	9,42 - 504	10,1 - 441	6,86 - 424	5,92 - 466	6,91 - 418	9,29 - 435
44	13,2 - 463	13,2 - 505	13,2 - 463	7,97 - 387	8,09 - 409	7,13 - 445	7,93 - 461	6,28 - 400	10,7 - 449	6,25 - 406	9,38 - 405	14,9 - 503
45	13,2 - 463	7,84 - 410	9,43 - 389	10,7 - 500	7,24 - 444	9,42 - 416	4,51 - 425	10,1 - 441	7,39 - 455	5,91 - 451	9,29 - 435	6,18 - 339
46	8,77 - 458	6,56 - 373	10,7 - 500	7,41 - 371	6,18 - 339	9,62 - 470	11,9 - 418	8,58 - 373	6,86 - 424	8,22 - 414	9,57 - 481	6,28 - 400
47	11,7 - 485	8,17 - 431	9,28 - 495	6,95 - 465	9,1 - 459	8,9 - 395	11 - 387	8,78 - 427	7,3 - 392	6,86 - 424	9,08 - 418	8,56 - 459
48	11,7 - 462	13,1 - 424	13,2 - 463	8,03 - 456	9,57 - 481	9,95 - 428	11 - 482	11,9 - 418	8,22 - 414	6,25 - 406	8,95 - 407	9,47 - 370
49	6,52 - 417	11 - 482	6,52 - 417	9,93 - 419	8,87 - 450	6,39 - 420	6,82 - 392	6,56 - 373	7,34 - 471	7,3 - 392	9,48 - 438	6,2 - 515
50	12,1 - 451	10 - 463	7,41 - 371	11,7 - 506	7,2 - 475	9,49 - 456	10 - 463	6,82 - 392	5,92 - 466	10,3 - 453	8,9 - 466	10,9 - 418
51	13,2 - 505	11,7 - 415	7,41 - 371	9,93 - 419	9,57 - 484	6,69 - 434	8,54 - 439	8,17 - 431	10,3 - 453	9 - 471	8,52 - 409	8,9 - 410
52	10,1 - 489	11,2 - 428	9,93 - 419	7,41 - 371	9,08 - 418	9,93 - 403	10 - 410	12,4 - 413	10,2 - 475	8,74 - 467	8,18 - 413	7,93 - 461
53	8,24 - 442	11,2 - 428	7,46 - 440	11,7 - 506	9,48 - 438	11,1 - 430	12,4 - 413	11,2 - 428	6,78 - 393	9,71 - 420	9,48 - 438	11,5 - 407
54	9,32 - 440	10 - 410	11,7 - 506	6 - 360	6,23 - 426	7,39 - 455	8,58 - 373	10 - 410	9,49 - 456	5,59 - 423	6,18 - 339	9,42 - 425
55	7,84 - 410	10,9 - 452	9,93 - 419	7,87 - 493	11 - 440	10,7 - 449	8,78 - 427	11 - 387	7,3 - 392	10,2 - 502	9,1 - 459	8,15 - 417
56	9,73 - 448	11 - 387	6,95 - 465	6,95 - 465	8,45 - 385	11,1 - 430	11 - 482	11 - 482	11,2 - 495	7,34 - 471	9,57 - 484	8,68 - 436

57	10,4 - 467	8,23 - 398	8,03 - 456	11,7 - 506	10,6 - 464	8,93 - 414	10 - 463	10 - 463	8,9 - 452	9 - 471	7,2 - 475	5,54 - 425
58	11,7 - 462	9,87 - 399	7,41 - 371	9,93 - 419	8,09 - 409	8,76 - 369	6,56 - 373	10,6 - 465	9,71 - 420	10,2 - 502	9,57 - 481	9,29 - 435
59	9,43 - 389	8,78 - 427	7,46 - 440	8,03 - 456	14,9 - 503	9,42 - 416	6,82 - 392	5,34 - 355	8,74 - 467	5,59 - 423	10,7 - 446	5,68 - 429
60	13,2 - 463	11,9 - 418	13,6 - 564	6,95 - 465	8,52 - 409	10,7 - 461	8,17 - 431	13,1 - 424	11,2 - 536	5,91 - 451	8,95 - 407	8,14 - 423
61	7,84 - 410	11,9 - 418	12,6 - 480	9,92 - 418	8,9 - 466	6,14 - 327	9,87 - 399	8,23 - 398	10,3 - 453	9,89 - 474	9,08 - 418	6,91 - 418
62	7,46 - 445	10 - 410	6 - 360	8,03 - 456	8,9 - 395	10,7 - 461	6,5 - 398	9,87 - 399	5,91 - 451	12,4 - 522	8,87 - 450	8,41 - 419
63	9,28 - 453	8,23 - 398	10,7 - 500	13,6 - 564	6,69 - 434	11,6 - 443	5,34 - 355	11,7 - 415	7,13 - 434	11,2 - 536	9,09 - 463	9,12 - 416
64	8,39 - 412	9,87 - 399	7,46 - 440	6 - 360	7,38 - 477	8,97 - 439	10,6 - 465	9,89 - 429	10,3 - 453	10,2 - 475	8,9 - 466	6,39 - 432
65	8,49 - 442	10,6 - 465	11,7 - 506	9,92 - 418	9,62 - 470	8,93 - 414	10,9 - 452	6,5 - 398	9,71 - 420	9,28 - 424	8,52 - 409	10,9 - 483
66	11,7 - 462	8,77 - 458	7,97 - 387	9,93 - 419	7,24 - 444	8,76 - 369	9,42 - 504	8,78 - 427	10,2 - 502	9,71 - 420	8,14 - 426	8,68 - 436
67	13,2 - 505	8,58 - 373	18,2 - 551	7,87 - 493	7,24 - 444	11,2 - 431	8,9 - 410	9,42 - 504	8,9 - 452	5,91 - 451	14,9 - 503	8,15 - 440
68	10,1 - 489	10,9 - 452	11,7 - 506	9,93 - 419	10,6 - 464	9,39 - 384	4,51 - 425	11 - 387	9 - 471	9,89 - 474	10,6 - 464	5,54 - 425
69	8,24 - 442	13,1 - 424	6 - 360	7,87 - 493	11 - 440	7,13 - 445	11,5 - 407	11,9 - 418	5,92 - 466	12,4 - 522	8,09 - 409	11,5 - 407
70	9,32 - 440	11 - 387	13,6 - 564	11,7 - 506	8,45 - 385	6,39 - 420	9,42 - 425	9,87 - 399	9,89 - 474	8,74 - 467	8,45 - 385	9,42 - 425
71	9,37 - 469	11,9 - 418	10,7 - 500	11,7 - 506	8,14 - 426	6,14 - 327	5,54 - 425	11,7 - 415	12,4 - 522	6,78 - 393	9,1 - 459	8,15 - 417
72	11,7 - 485	8,78 - 427	7,46 - 440	10,8 - 493	14,9 - 503	7,13 - 445	10,9 - 418	10 - 463	9 - 471	7,34 - 471	9,08 - 418	10,1 - 441
73	9,37 - 469	8,58 - 373	7,46 - 440	10,8 - 493	10,7 - 446	7,39 - 455	6,28 - 400	6,56 - 373	8,9 - 452	6,78 - 393	8,95 - 407	10,9 - 418
74	12,1 - 451	12,4 - 413	8,03 - 456	10,8 - 493	8,52 - 409	11,1 - 430	8,15 - 440	8,17 - 431	5,59 - 423	6,25 - 406	9,57 - 484	8,56 - 459
75	9,51 - 488	8,54 - 439	6,52 - 417	10,8 - 493	8,9 - 466	9,93 - 403	10,1 - 441	8,54 - 439	9,71 - 420	6,86 - 424	11 - 440	9,47 - 370
76	8,49 - 442	11,7 - 415	9,51 - 488	11 - 465	6,18 - 339	7,38 - 477	8,56 - 459	12,4 - 413	8,74 - 467	8,22 - 414	9,57 - 481	8,87 - 450
77	9,28 - 453	6,56 - 373	9,43 - 389	9,93 - 419	9,1 - 459	5,22 - 448	9,47 - 370	8,23 - 398	11,2 - 536	7,13 - 434	9,09 - 463	9,09 - 463
78	8,39 - 412	10 - 463	11 - 465	9,73 - 448	6,23 - 426	8,86 - 479	5,68 - 429	13,1 - 424	10,2 - 475	11,8 - 521	8,87 - 450	8,18 - 413
79	7,84 - 410	11 - 482	11 - 465	10,4 - 467	8,18 - 413	8,9 - 395	8,14 - 423	6,5 - 398	11,2 - 495	10,3 - 453	8,18 - 413	6,23 - 426
80	7,46 - 445	6,82 - 392	9,28 - 453	13,4 - 538	9,48 - 438	7,15 - 489	6,39 - 432	10,9 - 452	8,22 - 414	11,2 - 536	9,57 - 484	6,18 - 339
81	9,43 - 389	6,56 - 373	8,25 - 397	6 - 360	9,08 - 418	6,14 - 327	6,39 - 432	8,58 - 373	6,78 - 393	10,2 - 475	9,1 - 459	8,9 - 410
82	13,2 - 463	11 - 482	8,25 - 397	12,7 - 434	9,62 - 470	11,6 - 443	10,9 - 483	8,78 - 427	10,2 - 475	11,2 - 495	7,2 - 475	4,51 - 425
83	6,52 - 417	8,17 - 431	8,25 - 397	11 - 465	7,2 - 475	10,7 - 461	9,38 - 405	11,9 - 418	11,2 - 495	8,74 - 467	10,6 - 464	9,29 - 435
84	8,77 - 458	8,54 - 439	8,39 - 412	8,03 - 456	7,2 - 475	9,42 - 416	6,91 - 418	11 - 387	7,3 - 392	9,89 - 474	8,45 - 385	5,68 - 429
85	11,7 - 485	9,87 - 399	8,39 - 412	11 - 465	9,57 - 484	7,39 - 455	9,42 - 504	10 - 410	9,89 - 474	11,8 - 521	8,09 - 409	8,14 - 423
86	8,24 - 442	8,23 - 398	13,4 - 538	6,17 - 395	9,1 - 459	8,76 - 369	11,5 - 407	11 - 482	5,91 - 451	9 - 471	9,08 - 418	6,91 - 418
87	10,1 - 489	12,4 - 413	13,4 - 538	11 - 465	6,18 - 339	11,1 - 430	4,51 - 425	10 - 463	5,92 - 466	8,9 - 452	8,95 - 407	8,41 - 419
88	13,2 - 505	11,2 - 428	10,7 - 500	9,92 - 418	8,18 - 413	8,97 - 439	9,29 - 435	6,56 - 373	11,8 - 521	10,2 - 502	9,57 - 484	9,12 - 416
89	13,2 - 505	10 - 410	7,46 - 445	11 - 465	9,09 - 463	9,39 - 384	8,9 - 410	8,17 - 431	6,86 - 424	9,71 - 420	8,14 - 426	6,39 - 432
90	10,1 - 489	8,17 - 431	8,03 - 456	8,25 - 397	8,87 - 450	7,13 - 445	9,42 - 425	8,54 - 439	8,22 - 414	6,25 - 406	6,23 - 426	10,9 - 483
A	7,88 - 377	12,9 - 497	13,6 - 481	13,5 - 516	9,59 - 414	9,87 - 399	11,6 - 409	12,6 - 447	11,8 - 500	12,1 - 445	12,5 - 444	9,59 - 437
B	14,2 - 464	11,7 - 453	12 - 466	11,3 - 444	13,3 - 454	8,29 - 375	15,2 - 459	9,94 - 418	9,88 - 424	14,8 - 464	11,6 - 478	9,86 - 414
C	14,5 - 458	12,4 - 453	12,7 - 529	15,6 - 477	14,4 - 476	15,5 - 479	8,98 - 356	14,2 - 469	12,8 - 448	10,1 - 445	12,3 - 483	13,2 - 484
D	14,3 - 439	8,97 - 413	12,3 - 435	11,1 - 541	12,7 - 448	13,9 - 470	11,3 - 445	13,2 - 455	13,3 - 449	12,3 - 430	8,59 - 383	11,4 - 428
E	9,52 - 415	11,5 - 439	10,7 - 458	10,1 - 399	12 - 408	8,83 - 437	13,9 - 454	9,39 - 422	14,3 - 462	14,5 - 461	13,4 - 447	9,95 - 399
F	11,5 - 443	10,1 - 415	16,5 - 546	12,5 - 543	10,5 - 397	11,7 - 458	11,8 - 441	8,54 - 391	15,6 - 507	13,6 - 466	11,1 - 477	11,7 - 420
G	12,9 - 444	11,8 - 448	11,9 - 453	15,6 - 490	15,1 - 471	6,86 - 392	10,8 - 439	9,91 - 401	12,2 - xxx	13,1 - 488	13,7 - 445	11,6 - 463
H	13,3 - 459	13,7 - 529	12,4 - 458	13,2 - 511	11,1 - 410	11,1 - 434	8,3 - 403	12,8 - 505	10,8 - 439	7,13 - 404	9,64 - 415	14,9 - 525
I	13,6 - 458	13,2 - 536	11,9 - 451	15,1 - 508	13,4 - 479	8,84 - 395	10,7 - 438	14,5 - 476	12,8 - 480	13 - 453	9,46 - 400	14,8 - 532
J	11 - 490	12,3 - 529	9,76 - 483	10,8 - 391	10,6 - 489	10,9 - 418	12,2 - 439	13,3 - 481	12,8 - xxx	15 - 473	11 - 435	13,8 - 482
K	14,8 - 497	14,1 - 491	16,8 - 510	13,4 - 511	11,8 - 467	13,7 - 473	12,3 - 498	12,1 - 439	12,5 - xxx	12,3 - 498	12,2 - 481	9,41 - 442
L	12,6 - 428	13,6 - 467	10,5 - 410	15,8 - 528	11,5 - 443	11,2 - 468	13,6 - 491	13,9 - 460	13,1 - 460	15,7 - 452	15 - 511	12,3 - 424
M	11 - 434	13,2 - 402	12,4 - 435	14,1 - 450	8,56 - 347	11,7 - 449	10,7 - 441	12,3 - 513	12,3 - 473	14,7 - 480	13,1 - 462	11,6 - 367
N	7,55 - 401	11,4 - 361	9,49 - 438	11,7 - 472	11 - 429	11,2 - 409	15,8 - 477	11,3 - 450	12,4 - 453	9,22 - 407	13,4 - 488	12,6 - 471
O	12,9 - 464	16,2 - 446	15,8 - 475	15,7 - 446	13,3 - 500	13,5 - 453	10,7 - 417	11,3 - 462	12,9 - 520	13,7 - 464	9,65 - 535	12 - 429
P	12,5 - 481	14 - 490	11,1 - 406	10,2 - 421	13,6 - 491	16,2 - 494	16,1 - 514	13,3 - 499	13,4 - 459	14,1 - 461	11,3 - 413	12 - 486
Q	15 - 505	14,3 - 490	10,5 - 492	13,8 - 400	9,96 - 415	13,3 - 454	11,5 - 446	13,3 - 501	7,79 - 388	14,7 - 475	11,8 - 452	8,57 - 349
R	13,3 - 468	9,69 - 430	12,4 - 489	11,2 - 418	13,5 - 473	14,8 - 479	13,2 - 500	10,2 - 405	13,6 - 454	12,4 - 431	9,83 - 395	10,6 - 433

Table 16   Allocation of parameters in species 2-1, numeration displayed in Fig. 49

board \ beam	Dynamic MOE in kN/mm² - Dry timber density in kg/m³							
	1	2	3	4	5	6	7	8
layer orientation	/\	\/	\/	/\	/\	\/	\/	/\
1	14,8 - 512	9,5 - 366	8,95 - 399	11,3 - 456	8,28 - 428	8,53 - 441	9,97 - 419	15 - 470
2	6,23 - 426	7,45 - 444	10,8 - 472	10,2 - 495	11,5 - 445	12,4 - 465	11 - 449	9,32 - 450
3	15,7 - 539	11,7 - 478	12,4 - 473	11,2 - 495	11,5 - 445	10,7 - 410	10,5 - 467	11,2 - 394
4	12 - 405	10,6 - 475	8,56 - 431	11,2 - 495	11,5 - 445	10,7 - 410	12 - 405	11,2 - 394
5	9,89 - 472	11,7 - 478	8,56 - 431	9,71 - 467	11,5 - 445	10,7 - 410	7,45 - 444	10,4 - 463
6	9,32 - 450	11,1 - 464	7,47 - 402	9,71 - 467	7,29 - 366	10,7 - 410	13,3 - 502	10,4 - 463
7	12,1 - 470	7,45 - 444	7,47 - 402	11,2 - 495	7,29 - 366	9,98 - 493	7,56 - 410	10,5 - 461
8	9,83 - 404	13,6 - 468	8,56 - 431	8,52 - 464	6,49 - 424	9,98 - 493	11,2 - 438	10,5 - 461
9	15,7 - 539	11,9 - 440	8,56 - 431	8,52 - 464	6,49 - 424	10,9 - 406	11,2 - 438	12,9 - 407
10	10,4 - 463	10,6 - 475	9,54 - 405	8,52 - 464	6,49 - 424	15,3 - 540	11,8 - 502	12,9 - 407
11	11,4 - 448	10,5 - 450	9,54 - 405	12,2 - 466	6,49 - 424	15,3 - 540	11,8 - 502	9,32 - 448
12	11,2 - 390	11,7 - 460	9,54 - 405	12,2 - 466	7,29 - 366	9,98 - 493	10,2 - 480	9,32 - 448
13	13,9 - 462	11,1 - 464	7,47 - 402	8,52 - 464	7,29 - 366	10,7 - 418	10,2 - 480	9,93 - 433
14	12,5 - 455	10,6 - 475	7,47 - 402	12,2 - 466	12 - 501	15,3 - 540	11,4 - 448	9,93 - 433
15	7,56 - 410	11,7 - 478	12 - 501	12,2 - 466	12 - 501	15,3 - 540	10,8 - 466	9,71 - 467
16	13,6 - 468	10,9 - 464	10,4 - 462	13,1 - 487	12 - 501	10,8 - 471	10,8 - 466	9,71 - 467
17	13,6 - 468	10,9 - 464	10,4 - 462	13,1 - 487	9,96 - 419	10,8 - 471	10,4 - 441	10,2 - 495
18	7,56 - 410	11,7 - 478	10,4 - 462	11,1 - 457	9,96 - 419	10,3 - 440	10,4 - 441	10,2 - 495
19	11,9 - 440	11,7 - 460	10,4 - 462	13,1 - 487	9,96 - 419	10,3 - 440	10,1 - 429	10,3 - 440
20	11,9 - 440	7,45 - 444	10,8 - 437	11,1 - 457	9,96 - 419	11,2 - 394	10,1 - 429	10,3 - 440
21	12,5 - 480	10,5 - 440	11,5 - 380	9,32 - 450	13,1 - 526	14 - 469	9,81 - 430	9,52 - 404
22	9,11 - 429	10,1 - 423	10,9 - 454	11,9 - 483	8,6 - 415	12,2 - 495	7,35 - 433	11,5 - 468
23	13,9 - 462	10,1 - 429	10,8 - 472	13,1 - 487	11,6 - 491	12,4 - 465	12,5 - 455	9,32 - 450
24	11,2 - 390	10,4 - 441	10,8 - 472	10,8 - 437	10,7 - 380	10,2 - 440	12,5 - 455	9,32 - 450
25	11,4 - 448	10,8 - 466	6,23 - 361	12 - 475	10,7 - 380	10,2 - 440	12,5 - 455	14 - 584
26	10,8 - 466	11,4 - 448	6,23 - 361	10,9 - 406	10,7 - 380	9,98 - 493	13,3 - 502	14 - 584
27	10,4 - 441	8,08 - 413	6,23 - 361	10,9 - 406	10,7 - 380	9,98 - 493	10,1 - 452	14 - 584
28	10,1 - 429	11,2 - 390	12,2 - 461	10,9 - 406	10,2 - 523	10,2 - 440	11,7 - 460	14 - 584
29	13,6 - 468	13,9 - 462	12,2 - 461	12 - 475	10,2 - 523	10,2 - 440	9,83 - 404	9,89 - 472
30	12,6 - 530	11 - 449	12,2 - 461	12 - 475	12,7 - 458	10,5 - 461	13,9 - 462	9,89 - 472
31	10,1 - 452	9,83 - 404	7,56 - 444	12 - 475	12,7 - 458	10,5 - 461	11,2 - 390	9,89 - 472
32	9,83 - 404	12,1 - 470	7,56 - 444	9,53 - 468	11,6 - 491	12,4 - 465	8,08 - 413	10,4 - 432
33	10,1 - 452	12,6 - 530	7,56 - 444	10,8 - 437	11,6 - 491	12,4 - 465	8,08 - 413	10,4 - 432
34	10,5 - 450	13,3 - 502	7,56 - 444	10,8 - 437	11,6 - 491	9,32 - 448	8,08 - 413	10,4 - 432
35	10,5 - 450	12 - 442	10,5 - 450	11,1 - 457	10,2 - 523	9,32 - 448	15,8 - 507	11,2 - 438
30	13,3 - 502	7,66 - 410	14,1 - 507	11,1 - 457	10,2 - 523	9,93 - 433	15,8 - 507	11,2 - 438
37	12,6 - 530	11,9 - 440	14,1 - 507	9,53 - 468	12,7 - 458	9,93 - 433	15,8 - 507	11,8 - 502
38	12,1 - 470	12 - 405	14,1 - 507	9,53 - 468	12,7 - 458	11,2 - 394	15,8 - 507	11,8 - 502
39	11 - 449	15,7 - 539	14,1 - 507	9,53 - 468	12,4 - 473	10,8 - 472	10,5 - 467	10,2 - 480
40	11 - 449	12 - 405	12,2 - 461	12,9 - 407	12,4 - 473	6,23 - 361	10,5 - 467	10,2 - 480
41	15,7 - 539	10,1 - 452	12,4 - 473	12,9 - 407	11,2 - 495	9,54 - 405	10,5 - 467	10,4 - 463
42	10,7 - 417	8,73 - 423	9,57 - 464	8,19 - 430	9,22 - 398	10,1 - 441	10,1 - 457	9,37 - 431
A	11,7 - 443	11 - 403	14,7 - 464	11,7 - 432	11 - 446	12,3 - 449	11,7 - 460	9,01 - 399
B	14,1 - 472	9,71 - 436	11,8 - 459	12,3 - 428	12,8 - 450	15 - 474	12,7 - 468	13,3 - 498
C	15,6 - 491	11,1 - 419	11,6 - 466	11,1 - 418	10,2 - 452	8,64 - 437	10,8 - 456	9,88 - 429
D	11,8 - 387	13,7 - 454	12,5 - 470	16,2 - 524	11,5 - 427	7,88 - 418	12,4 - 514	10,3 - 419
E	15,5 - 489	10,6 - 349	9,63 - 428	9,26 - 357	11,9 - 420	10,3 - 382	11,8 - 400	17,4 - 516
F	13,8 - 471	11 - 433	7,78 - 425	9,63 - 399	12,3 - 486	11,4 - 452	10,7 - 431	11,4 - 419
G	10,6 - 428	13,6 - 491	14,4 - 468	8,8 - 408	10,3 - 454	11 - 445	14,3 - 449	11 - 470
H	10,8 - 377	10,1 - 351	11,5 - 431	9,29 - 414	11,9 - 445	16,4 - 527	13,8 - 480	14,3 - 489
I	7,76 - 401	12,5 - 493	10,8 - 431	15,8 - 522	12,9 - 451	13 - 458	12,3 - 480	12,6 - 428
J	11,5 - 394	11 - 434	14,1 - 518	13 - 465	14 - 483	10,3 - 442	9,89 - 408	9,85 - 414
K	9,78 - 390	14,2 - 498	10,2 - 398	9,44 - 410	8,97 - 391	10,7 - 453	11,3 - 445	8,94 - 407
L	13,4 - 447	7,98 - 404	12,7 - 457	12,5 - 482	13,7 - 473	13,4 - 466	9,49 - 443	10,8 - 468

Table 17   Allocation of parameters in species 2-2, numeration displayed in Fig. 50

board / beam	Dynamic MOE in kN/mm² - Dry timber density in kg/m³									
	1	2	3	4	5	6	7	8	9	10
layer orientation	/ \	\ /	/ \	\ /	\ /	/ \	\ /	/ \	/ \	/ \
1	11,9 - 474	7,16 - 400	11,5 - 438	9,56 - 429	11,6 - 437	9,96 - 447	12,6 - 466	9,56 - 448	9,33 - 456	16,6 - 497
2	9,52 - 475	16,5 - 526	10,2 - 439	9,55 - 488	9,74 - 451	12,2 - 429	9,22 - 398	13,7 - 438	13,1 - 532	8,6 - 452
3	11,4 - 424	12,5 - 477	9,5 - 428	9,55 - 488	11,1 - 390	10,9 - 454	9,55 - 499	13,7 - 438	10,9 - 464	14 - 584
4	13,1 - 526	10,2 - 439	10,5 - 473	12,6 - 495	10,5 - 440	10,9 - 454	12,2 - 429	9,18 - 464	8,47 - 437	9,04 - 447
5	9,94 - 369	10,5 - 457	8,24 - 427	10 - 472	11,5 - 419	10,8 - 387	8,52 - 464	6,86 - 452	12 - 442	8,6 - 415
6	10,2 - 439	10,3 - 439	11,4 - 476	10,5 - 457	10,8 - 406	10,8 - 387	11,4 - 530	6,86 - 452	11,7 - 460	8,6 - 415
7	11,7 - 450	12,5 - 477	13,3 - 452	13,1 - 526	13,3 - 452	9,84 - 462	10,3 - 512	8,52 - 464	10,5 - 450	7,12 - 462
8	9,54 - 444	10 - 472	11,7 - 450	13,1 - 526	8,86 - 403	9,84 - 462	10,3 - 512	8,52 - 464	8,47 - 437	12 - 442
9	11,6 - 401	12,6 - 495	9,45 - 473	11,2 - 398	11,4 - 476	13,5 - 530	11,4 - 530	8,28 - 428	7,12 - 462	11,1 - 460
10	9,5 - 428	9,5 - 428	14,8 - 514	12,5 - 502	11,7 - 450	13,5 - 530	8,52 - 464	8,28 - 428	10,6 - 475	13,1 - 532
11	11,5 - 419	9,94 - 369	8,97 - 424	11,2 - 398	9,45 - 473	8,6 - 415	12,2 - 429	8,28 - 428	8,47 - 437	10,2 - 412
12	12,8 - 485	10,1 - 457	9,92 - 495	12,5 - 502	14,8 - 514	8,6 - 415	9,55 - 499	8,53 - 441	7,45 - 423	9,13 - 414
13	13,3 - 501	10,4 - 473	11,6 - 401	9,5 - 366	9,84 - 462	12,2 - 458	9,22 - 398	11,4 - 530	9,04 - 447	11,1 - 464
14	12,8 - 485	12,5 - 477	9,52 - 475	12,6 - 495	13,5 - 530	12,2 - 458	8,95 - 399	11,4 - 530	10,9 - 464	10,2 - 412
15	9,54 - 444	9,31 - 458	7,67 - 471	16,5 - 526	10,1 - 441	7,98 - 439	8,26 - 468	10,3 - 512	11,1 - 460	9,13 - 414
16	9,92 - 495	10,3 - 439	13,3 - 501	9,5 - 366	11,1 - 467	7,98 - 439	11,1 - 467	10,3 - 512	11,1 - 460	9,13 - 414
17	12,5 - 480	10,3 - 439	12,8 - 485	10,1 - 457	8,53 - 441	14,5 - 496	10,1 - 441	6,89 - 406	12 - 442	9,13 - 414
18	11,7 - 450	15,5 - 506	9,94 - 369	12,1 - 446	15,6 - 518	14,5 - 496	12 - 465	6,89 - 406	13,1 - 532	10,2 - 412
19	9,52 - 475	15,5 - 506	10,8 - 406	9,5 - 366	6,89 - 406	14,5 - 496	8,53 - 441	13,3 - 452	13,1 - 532	10,2 - 412
20	7,67 - 471	10 - 472	11,5 - 419	8,73 - 423	8,24 - 427	14,5 - 496	15,6 - 518	12 - 465	11,1 - 460	11,1 - 464
21	8,91 - 447	9,36 - 423	8,68 - 442	10,1 - 437	14,4 - 528	9,92 - 495	11,2 - 453	10,5 - 462	12,2 - 465	12,8 - 460
22	11,5 - 428	11,1 - 460	11,2 - 459	9,65 - 496	7,95 - 442	8,97 - 424	11,3 - 460	11,1 - 450	8,97 - 447	8,88 - 446
23	7,67 - 471	10,1 - 457	11,5 - 468	12,1 - 446	11,1 - 390	10,1 - 441	9,96 - 447	9,55 - 499	13,7 - 438	7,45 - 423
24	11,1 - 390	8,24 - 427	10,8 - 406	8,73 - 423	11,1 - 390	9,22 - 398	9,74 - 451	15 - 470	9,03 - 413	7,12 - 462
25	9,74 - 451	11,4 - 476	10,5 - 440	10,4 - 473	11,4 - 424	11,1 - 467	10,9 - 454	9,06 - 471	9,64 - 439	15 - 470
26	12,8 - 485	8,73 - 423	10,5 - 440	11,7 - 493	10,5 - 473	9,31 - 458	8,8 - 456	10,1 - 441	10,7 - 448	9,04 - 447
27	12,2 - 495	10,4 - 473	12,5 - 480	9,55 - 488	9,74 - 451	11,1 - 467	10,8 - 387	12,2 - 429	9,03 - 413	15 - 470
28	9,88 - 492	11,4 - 424	8,86 - 403	15,5 - 506	13,3 - 452	9,22 - 398	13,5 - 530	8,26 - 468	9,03 - 413	11,7 - 493
29	9,54 - 444	10,2 - 439	10,1 - 423	12,1 - 446	10,8 - 406	8,8 - 466	10,1 - 423	8,95 - 399	12,2 - 458	9,18 - 464
30	8,24 - 427	9,5 - 428	8,86 - 403	12,5 - 502	9,96 - 447	14 - 584	6,86 - 452	8,8 - 456	7,98 - 439	8,6 - 452
31	14,8 - 514	9,94 - 369	9,88 - 492	11,2 - 398	11,5 - 468	12 - 465	10,8 - 387	8,26 - 468	10,7 - 448	9,31 - 458
32	9,45 - 473	11,2 - 398	11,6 - 401	14,8 - 514	12,5 - 480	8,28 - 428	6,89 - 406	7,99 - 424	9,06 - 471	7,12 - 462
33	8,97 - 424	15,5 - 506	12,2 - 495	9,92 - 495	11,5 - 468	9,06 - 471	9,84 - 462	8,8 - 456	10,7 - 448	9,03 - 413
34	9,92 - 495	9,55 - 488	9,52 - 475	10,5 - 473	11,5 - 419	13,7 - 438	8,85 - 442	8,85 - 442	7,98 - 439	8,47 - 437
35	10,5 - 473	12,6 - 495	7,67 - 471	10,1 - 457	9,96 - 447	9,55 - 499	12,5 - 480	7,99 - 424	12,2 - 458	9,31 - 458
36	11,4 - 476	16,5 - 526	9,88 - 492	9,5 - 366	11,5 - 468	15 - 470	10,1 - 423	13,3 - 501	8,8 - 466	9,04 - 447
37	11,6 - 401	12,1 - 446	12,2 - 495	10,3 - 439	8,86 - 403	12 - 465	7,99 - 424	8,85 - 442	9,31 - 458	11,7 - 493
38	9,45 - 473	10 - 472	11,4 - 424	10,5 - 457	10,9 - 454	8,8 - 466	8,8 - 456	8,53 - 441	9,31 - 458	7,45 - 423
39	9,88 - 492	8,73 - 423	8,97 - 424	12,5 - 502	7,99 - 424	9,18 - 464	8,26 - 468	15,6 - 518	10,7 - 448	8,6 - 452
40	12,2 - 495	13,1 - 526	9,96 - 447	12,5 - 477	8,95 - 399	11,7 - 493	10,5 - 440	15,6 - 518	8,6 - 452	9,18 - 464
41	13,3 - 501	8,97 - 424	10,1 - 423	10,4 - 473	6,86 - 452	8,95 - 399	8,85 - 442	9,54 - 444	9,31 - 458	9,31 - 458
42	11 - 465	9,94 - 405	9,27 - 445	15 - 488	11,4 - 424	14,8 - 514	10,2 - 463	9,58 - 456	11,9 - 438	13,5 - 471
A	7,45 - 393	11,2 - 453	10,9 - 458	12,7 - 508	14,9 - 499	14,7 - 459	11,9 - 472	10,5 - 400	10,9 - 458	11,9 - 449
B	12,3 - 480	13 - 421	11,2 - 437	12,9 - 461	12,2 - 435	11,3 - 451	11,1 - 433	12,9 - 449	17,1 - 568	8,24 - 403
C	14,6 - 469	8,08 - 400	11,3 - xxx	12,6 - 465	12,4 - 491	14,7 - 502	11,9 - 473	10,1 - 427	12,4 - 406	11,2 - 419
D	13,5 - 452	10,6 - 414	11,9 - xxx	12,7 - 423	8,6 - 416	12,2 - 535	11,2 - 449	10,6 - 426	9,02 - 397	12,3 - 401
E	6,31 - 392	15,5 - 448	11,2 - 423	11 - 449	8,89 - 422	7,01 - 381	7,58 - 380	10,4 - 425	14,9 - 508	10,4 - 430
F	11,3 - 380	11,9 - 476	9,9 - 408	11,5 - 442	11,2 - 425	12,8 - 398	13,8 - 473	14,6 - 515	9,34 - 391	13,7 - 415
G	10,8 - 433	9,78 - 434	11 - 426	10,3 - 416	14,7 - 541	12,9 - 596	12,5 - 486	11,2 - 435	13 - 458	10,1 - 398
H	12,4 - 498	9,71 - 405	15,6 - 535	12,6 - 484	12,6 - 497	8,93 - 418	15,2 - 452	11,6 - 439	11,7 - 378	13,8 - 460

Table 18   Allocation of parameters in species 2-3, numeration displayed in Fig. 51

board \ beam	Dynamic MOE in kN/mm^2 - Dry timber density in kg/m^3							
	1	2	3	4	5	6	7	8
layer orientation	/\	\/	\/	/\	/\	\/	\/	/\
1	11,7 - 460	13,3 - 502	9,85 - 461	15,2 - 523	9,03 - 413	11,5 - 459	10,7 - 448	7,45 - 444
2	6,07 - 387	9 - 420	12,1 - 484	9,06 - 419	11,4 - 433	9,3 - 469	12,2 - 394	10,2 - 480
3	7,36 - 386	10,2 - 480	9,06 - 419	14,6 - 445	7,62 - 429	7,02 - 438	11,9 - 450	8,66 - 429
4	8,26 - 414	9,81 - 486	7,58 - 496	7,58 - 496	13,8 - 456	16,1 - 514	8,16 - 447	11,8 - 435
5	8,58 - 434	10,1 - 449	14,6 - 445	10,8 - 444	10,7 - 435	11 - 462	9,83 - 434	10,3 - 403
6	9,58 - 456	8,93 - 440	10,8 - 444	6,48 - 428	11 - 462	11,3 - 444	11,8 - 493	11,9 - 450
7	11,1 - 450	7,18 - 436	10,2 - 439	7,98 - 510	9,3 - 469	9,31 - 361	10,1 - 449	6,54 - 411
8	9,19 - 415	12,2 - 394	8,26 - 454	10,2 - 439	7,36 - 386	6,48 - 428	7,77 - 395	9,57 - 526
9	7,4 - 384	9,68 - 454	11,5 - 439	9,37 - 431	11,4 - 433	11,8 - 435	7,77 - 371	11,5 - 428
10	5,56 - 405	9,19 - 415	11,3 - 456	7,35 - 433	13,5 - 484	13,8 - 456	9,85 - 416	10,2 - 480
11	7,49 - 447	8,58 - 434	12,7 - 461	10 - 429	9,05 - 408	12,2 - 462	10,1 - 403	7,77 - 395
12	9,68 - 454	7,77 - 371	9,37 - 431	12,1 - 484	12,2 - 462	12 - 415	8,93 - 440	9,05 - 408
13	7,49 - 447	11,3 - 444	10 - 429	12,7 - 461	12 - 415	12 - 415	10,8 - 447	7,77 - 395
14	7,77 - 371	10,1 - 403	14,8 - 512	9,52 - 416	8,71 - 440	8,71 - 440	11,9 - 450	6,48 - 428
15	11,5 - 434	8,52 - 494	10,2 - 487	8,44 - 428	10,3 - 445	9,76 - 472	8,99 - 384	8,36 - 481
16	7,98 - 510	9,83 - 434	8,26 - 454	8,3 - 360	9,35 - 437	8,99 - 391	12,2 - 394	9,57 - 526
17	7,77 - 395	14 - 530	10,2 - 439	10,8 - 447	10,5 - 459	14,9 - 492	8,16 - 447	11,5 - 428
18	8,66 - 429	8,66 - 429	10,1 - 426	10,8 - 447	11,8 - 435	10,7 - 435	11,8 - 493	9,83 - 435
19	8,26 - 414	15,2 - 485	9,11 - 429	8,3 - 360	8,19 - 412	11 - 462	8,26 - 414	6,15 - 408
20	11 - 517	8,99 - 391	11,7 - 462	8,44 - 428	9,31 - 361	9,3 - 469	10,1 - 449	8,91 - 396
21	12 - 401	10,1 - 449	10 - 428	9,52 - 416	11,3 - 444	11,4 - 433	11,7 - 433	9,83 - 435
22	9,56 - 429	10,1 - 403	10,2 - 487	11,1 - 450	13,5 - 484	8,76 - 402	9,83 - 434	7,77 - 371
23	13,2 - 447	11,9 - 450	10,5 - 450	11,5 - 434	8,13 - 466	14,9 - 492	13,5 - 484	6,52 - 432
24	10 - 428	9,83 - 434	10,7 - 417	9,97 - 419	11,5 - 428	7,44 - 420	8,3 - 360	9,19 - 415
25	8,3 - 360	8,16 - 447	13,2 - 447	10,1 - 426	14,4 - 535	9,02 - 422	8,44 - 428	11,8 - 435
26	11 - 517	9,68 - 454	9,06 - 419	11,3 - 456	11,5 - 382	7,62 - 412	14,4 - 535	9,81 - 486
27	7,65 - 411	7,49 - 447	7,58 - 496	11,3 - 456	11 - 462	9,43 - 395	11,3 - 444	9,05 - 408
28	7,18 - 436	8,58 - 434	14,6 - 445	10,5 - 450	10,7 - 435	9,76 - 472	9,31 - 361	10,3 - 403
29	6,52 - 432	10,1 - 449	9,97 - 419	11,5 - 439	7,62 - 412	6,07 - 387	11 - 517	12,2 - 394
30	12 - 401	11,7 - 433	13,2 - 447	12,7 - 461	9,02 - 422	7,97 - 387	9,22 - 410	12 - 401
31	10,8 - 447	8,52 - 494	10,1 - 426	11,5 - 439	9,26 - 453	9,76 - 472	7,65 - 411	7,77 - 395
32	8,44 - 428	10,1 - 403	7,35 - 433	10,5 - 450	14,9 - 492	8,76 - 402	7,18 - 436	8,19 - 412
33	9,52 - 416	12,2 - 394	9,76 - 448	12,1 - 484	7,44 - 420	11 - 462	8,52 - 494	
34	12,7 - 460	8,93 - 440	15,6 - 518	10 - 455	7,12 - 462	12,7 - 443	13,1 - 532	11,9 - 440
35	12,2 - 394	10,5 - 450	8,67 - 425	11,4 - 458	7,31 - 424	12 - 442	11,2 - 394	
36	9,63 - 461	9,56 - 429	7,98 - 510	8,26 - 454	5,56 - 405		6,15 - 408	8,66 - 388
37	7,62 - 429	6,21 - 441	13,2 - 447	10,2 - 487	7,36 - 386	11,5 - 428	9,83 - 435	8,91 - 396
38	9,56 - 448	9 - 420	10,1 - 426	9,37 - 431	5,56 - 405	13,5 - 484	8,99 - 384	13,5 - 484
39	10,6 - 444	9,16 - 437	10 - 428	12,7 - 461	9,76 - 472	9,35 - 437	10,8 - 447	8,99 - 384
40	10,7 - 417	6,07 - 387	9,37 - 431	14,6 - 445	9,57 - 526	10,5 - 459	10,1 - 403	5,62 - 416
41	15,6 - 518	14 - 530	10 - 429	7,58 - 496	14,4 - 535	11,5 - 382	8,93 - 440	10,2 - 480
42	7,18 - 436	7,36 - 386	7,35 - 433	9,06 - 419	9 - 420	8,76 - 402	8,44 - 428	5,62 - 416
43	8,26 - 414	8,93 - 440	9,97 - 419	10,2 - 439	9,85 - 416	7,02 - 438	8,3 - 360	9,85 - 416
44	11 - 517	8,93 - 440	9,63 - 461	10,2 - 439	11,3 - 444	7,02 - 438	12 - 401	8,78 - 429
45	7,62 - 429	14 - 530	9,58 - 456	7,98 - 510	9,31 - 361	13,8 - 456	9,56 - 429	10,5 - 459
46	7,02 - 438	9,31 - 361	11,1 - 450	7,98 - 510	8,19 - 412	6,48 - 428	8,66 - 388	10,3 - 403
47	9,11 - 429	7,36 - 386	11,5 - 434	10,8 - 444	11,8 - 435	13,8 - 456	6,54 - 411	5,62 - 416
48	11,7 - 462	6,15 - 408	9,56 - 448	7,58 - 496	10,5 - 459	12,2 - 462	11,7 - 433	14 - 530
49	14,8 - 512	9,19 - 415	9,11 - 429	9,06 - 419	9,35 - 437	12 - 415	8,26 - 414	8,58 - 434
50	6,52 - 432	8,58 - 434	10,7 - 417	14,6 - 445	9,05 - 408	8,71 - 440	14 - 530	8,78 - 429
51	11,8 - 493	5,62 - 416	10,6 - 444	10,8 - 444	13,5 - 484	10,3 - 445	9,52 - 416	10,5 - 459
52	9,22 - 410	8,16 - 447	7,35 - 433	9,58 - 456	8,13 - 466	7,62 - 412	7,65 - 411	8,19 - 412
53	6,21 - 441	8,99 - 384	10 - 429	9,63 - 461	11,5 - 382	8,99 - 391	8,52 - 494	10,2 - 480
54	10,7 - 447	11,9 - 450	11,7 - 462	9,56 - 448	14,4 - 535	14,9 - 492	11 - 517	10,3 - 403

55	8,99 - 391	11,7 - 433	9,97 - 419	10,6 - 444	11,5 - 428	7,44 - 420	6,52 - 432	6,54 - 411
56	7,4 - 384	9,26 - 453	13,2 - 447	14,8 - 512	7,62 - 412	9,02 - 422	7,18 - 436	9,35 - 437
57	9,56 - 429	5,56 - 405	9,11 - 429	11,7 - 462	9,02 - 422	9,43 - 395	8,13 - 466	8,16 - 447
58	6,52 - 432	5,56 - 405	10,1 - 426	9,11 - 429	9,19 - 415	9,43 - 395	8,78 - 429	8,78 - 429
59	7,65 - 411	8,36 - 481	10 - 428	10,7 - 417	14,9 - 492	7,62 - 412	14,4 - 535	8,91 - 396
60	9,22 - 410	8,99 - 384	9,76 - 448	10 - 428	8,76 - 402	9,3 - 469	9,22 - 410	8,66 - 429
61	16,1 - 514	11,8 - 493	9,56 - 448	11,7 - 462	9,43 - 395	7,44 - 420	9,68 - 454	9,56 - 429
62	9 - 420	11,7 - 433	10,6 - 444	10,8 - 444	9,76 - 472	10,7 - 435	7,49 - 447	9,68 - 454
63	6,07 - 387	7,4 - 384	10,7 - 417	14,8 - 512	10,3 - 445	9,16 - 437	11,5 - 382	8,36 - 481
64	9,83 - 434	7,62 - 429	9,58 - 456	12,1 - 484	8,71 - 440	12,2 - 462	11,5 - 382	7,49 - 447
65	8,52 - 494	9,59 - 443	11,1 - 450	11,5 - 439	12 - 415	9,3 - 469	8,13 - 466	9,35 - 437
66	11,8 - 493	8,99 - 391	9,63 - 461	10,5 - 450	12,2 - 462	11,4 - 433	9,81 - 486	9,83 - 435
67	7,65 - 411	8,66 - 429	11,5 - 434	11,3 - 456	13,8 - 456	10,7 - 435	9,81 - 486	8,36 - 481
68	14 - 530		15 - 524	10,7 - 459	9,04 - 447	11,8 - 465	11,4 - 530	12 - 405
A	9,58 - 407	13,3 - 469	14,4 - 475	13,9 - 458	9,94 - 436	12,4 - 452	15,1 - 472	16,3 - 501
B	8,15 - 428	10,8 - 448	11,7 - 497	12,3 - 441	10,3 - 441	9,26 - 424	10,7 - 447	10,5 - 421
C	14,1 - 476	8,75 - 431	9,78 - 505	8 - 378	12,1 - 439	10,7 - 430	12,6 - 450	13,9 - 494
D	18,2 - 566	15 - 475	15,7 - 545	13 - 496	9,95 - 400	7,96 - 399	12 - xxx	13,2 - 465
E	13,3 - 420	13,7 - 501	15,1 - 481	12,6 - 481	11,9 - 456	10,2 - 422	9,11 - 395	14,8 - 486
F	12,4 - 427	10,3 - 430	10,8 - 395	10,3 - 425	13,8 - 500	9,58 - 390	12,7 - 440	14,7 - 488
G	10,5 - 428	10,3 - 439	11,4 - 417	13,2 - 459	9,87 - 416	10,8 - 459	13,8 - 510	11,4 - 448
H	14,7 - 530	11,6 - 403	11,6 - 475	9,1 - 418	7,34 - 373	13,9 - 535	7,38 - 396	11,9 - 429
I	12,3 - 449	10,1 - 441	9,95 - 399	13,3 - 475	14,4 - 474	10,4 - 458	13,4 - 447	12,5 - 463
J	14,5 - 448	15,1 - 463	13,2 - 475	9,69 - 408	11,5 - 434	12,9 - 455	13,9 - 440	13,2 - 450
K	16,6 - 537	13,6 - 480	11,2 - 465	11,4 - 389	13,5 - 483	11,2 - 399	10,7 - 473	10,6 - 422
L	9,45 - 408	13,8 - 501	12 - 446	7,62 - 367	10,9 - 454	13,8 - 469	12,6 - 459	10,5 - 334
M	11,7 - 478	12,9 - 505	16,3 - 577	12,2 - 513	10,9 - 476	13,3 - 488	12,8 - 456	8,48 - 393
N	6,67 - 427	13,6 - 467	14 - 471	9,29 - 404	9,17 - 422	10,6 - 488	9,2 - 404	16,4 - 477
O	11,1 - 417	11,1 - 448	9,12 - 408	10,8 - 419	13,1 - 443	13,5 - 440	14,3 - 464	10 - 438
P	9,03 - 432	12,4 - 408	13,8 - 475	13,1 - 481	10,7 - 399	9,58 - 418	12,6 - 444	12,9 - 522
Q	11,7 - 429	12,2 - 494	11,7 - 393	14,3 - 515	13,2 - 542	10,6 - 388	12,1 - 474	14,9 - 565
R	11,2 - 434	14,1 - 499	12,9 - 444	11,6 - 420	12,4 - 422	12,5 - 451	11,2 - 421	13 - 462

Table 19   Allocation of parameters in species 2-4, numeration displayed in Fig. 52

board \ beam	Dynamic MOE in kN/mm^2 - Dry timber density in kg/m^3									
	1	2	3	4	5	6	7	8	9	10
layer orientation	/\	\/	/\	\/	\/	/\	\/	/\	/\	/\
1	8,39 - 429	13,9 - 486		8,86 - 403	11,5 - 445	10,3 - 440	10,9 - 406	9,13 - 414	8,16 - 447	12 - 405
2	14 - 584	11 - 465		9,33 - 456	8,19 - 430	11,2 - 451	5,46 - 409	15,2 - 523	15,5 - 543	8,92 - 438
3	7,97 - 387	13,6 - 475	11,3 - 456	12,8 - 460	12,2 - 440	9,65 - 496	12,2 - 555	15 - 524	12,2 - 439	11 - 453
4	12,6 - 466	11,5 - 438	14,8 - 512	9,52 - 404	13,5 - 477	9,85 - 461	13,9 - 486	7,56 - 444	9,64 - 439	8,97 - 447
5	14 - 469	10,3 - 488	14,8 - 498	9,67 - 410	7,91 - 439	12,9 - 435	9,53 - 437	9,94 - 405	7,4 - 384	8,38 - 433
6	12,2 - 465	12,7 - 474	10,9 - 418	8,92 - 438	16,8 - 496	6,48 - 400	8,88 - 446	11 - 453	6,21 - 441	9,32 - 450
7	16,8 - 496	9,21 - 453	9,36 - 423	10,8 - 455	12,6 - 466	13,4 - 487	13 - 507	7,12 - 443	9,22 - 410	7,12 - 443
8	14,4 - 528	9,2 - 499	9,27 - 445	11,6 - 434	11,2 - 453	11,9 - 438	12 - 477	8,38 - 433	10,1 - 497	10,4 - 469
9	8,19 - 430	10,7 - 498	9,76 - 448	6,62 - 429	14,4 - 528	10 - 455	10,5 - 424	10,4 - 469	7,97 - 448	10,8 - 455
10	7,95 - 442	13 - 470	9,97 - 419	7,12 - 443	11,8 - 453	7,31 - 424	10 - 455	9,32 - 450	5,87 - 462	9,71 - 428
11	7,97 - 387	11,5 - 417	7,35 - 433	8,92 - 438	9,65 - 496	11,8 - 465	11,9 - 438	8,67 - 425	12,6 - 433	11,4 - 458
12	5,87 - 462	9,94 - 405	12,7 - 461	8,38 - 433	8,91 - 447	11,5 - 459	11,9 - 483	12,7 - 460	7,97 - 387	12 - 457
13	7,84 - 410	14,7 - 484	10,5 - 462	7,16 - 400	11,2 - 451	11,2 - 499	12,1 - 467	12,5 - 461	12,6 - 433	8,77 - 433
14	8,79 - 421	10,5 - 462	8,68 - 442	11 - 453	11,5 - 459	13,9 - 486	9,57 - 464	12,5 - 453	9,16 - 437	13,8 - 471
15	8,79 - 421	8,68 - 442	13 - 470	6,62 - 429	12,7 - 443	12,2 - 555	11,5 - 380	10,2 - 448	9,59 - 443	10,7 - 459
16	9,59 - 443	11,6 - 496	11,5 - 417	11,6 - 434	16,8 - 496	6,73 - 460	12,4 - 462	12,8 - 460	10,1 - 497	16,6 - 497
17	15,5 - 543	13 - 470	11,5 - 417	9,94 - 405	12,2 - 465	5,46 - 409	12,7 - 500	9,33 - 456	9,26 - 453	13,5 - 471
18	10 - 493	10,3 - 488	12,7 - 474	12,5 - 453	8,97 - 453	11,2 - 459	9,99 - 388	13,8 - 471	6,12 - 396	12,7 - 460
19	8,78 - 429	11,5 - 438	11,6 - 496	12,5 - 461	11,2 - 451	11,8 - 453	11,2 - 499	8,77 - 433	12,2 - 439	9,32 - 450
20	6,21 - 441	11,5 - 438	12,7 - 472	8,77 - 433	12,9 - 435	10,1 - 437	5,46 - 409	12 - 457	8,79 - 421	8,67 - 425
21	12,2 - 439	9,2 - 499	13,4 - 529	12 - 457	7,31 - 424	11,3 - 460	6,73 - 460	11,4 - 458	6,12 - 396	15,2 - 523
22	15,5 - 543	12,7 - 472	11,6 - 496	11,4 - 458	11,3 - 460	12,7 - 443	6,48 - 400	9,71 - 428	10,7 - 447	10,2 - 448
23	11,8 - 453	11 - 465	12,7 - 474	10,7 - 459	11,2 - 459	8,97 - 453	13,4 - 487	16,6 - 497	5,87 - 462	10,4 - 469
24	10,2 - 463	10,7 - 498	9,21 - 453	13,8 - 471	15 - 488	15 - 488	9,53 - 437	13,5 - 471	16,1 - 514	15 - 524
25	11,2 - 453	13,6 - 475	9,27 - 445	8,97 - 447	10,2 - 463	8,91 - 447	9,85 - 461	9,67 - 410	7,97 - 448	10,8 - 455
26	14,4 - 528	14,7 - 484	13 - 470	9,33 - 456	8,19 - 430	9,57 - 429	12,4 - 462	10,7 - 429	12,6 - 433	13,4 - 487
27	11,6 - 437	11,6 - 496	10,5 - 462	12,7 - 460	12,2 - 440	11,6 - 437	9,99 - 388	9,36 - 423	6,21 - 441	6,48 - 400
28	13,1 - 460	8,68 - 442	8,68 - 442	9,52 - 404	16,8 - 496	9,99 - 388	11,2 - 499	10,9 - 418	10,7 - 500	11,9 - 483
29	9,57 - 429	8,68 - 442	13,4 - 529	7,12 - 443	13,5 - 477	13 - 507	5,46 - 409	9,27 - 445	10,7 - 447	11,5 - 380
30	7,97 - 387	10,5 - 462	14,8 - 498	7,16 - 400	14 - 469	11,8 - 475	12,7 - 500	11,9 - 474	6,12 - 396	12,7 - 500
31	10 - 493	12,7 - 474	10,9 - 418	10,8 - 455	7,91 - 439	7,95 - 442	12,2 - 555	14,8 - 498	9,16 - 437	12,4 - 462
32	12,2 - 439	13,4 - 529	9,36 - 423	11,6 - 434	12,2 - 465	10,2 - 463	8,97 - 447	8,52 - 462	16,1 - 514	9,57 - 464
33	11,8 - 475	9,21 - 453	10,7 - 429	9,32 - 450	12,6 - 466	13,1 - 460	9,57 - 464	9,76 - 449	7,97 - 448	9,53 - 437
34	9,54 - 405	12,7 - 500	15,7 - 539	9,53 - 468	8,56 - 431	10,7 - 410	9,45 - 473	9,06 - 419	14,8 - 498	10,5 - 461
35	10,2 - 440	10,7 - 429	7,18 - 436	10,2 - 448	9,96 - 419	8,52 - 464	8,24 - 427	11,5 - 410	13,7 - 438	10,6 - 475
36	11,8 - 465	11,9 - 474	12,1 - 484	8,38 - 433	9,53 - 437	8,19 - 430	12,1 - 467	7,16 - 400	12,2 - 439	8,77 - 433
37	8,97 - 453	9,27 - 445	11,5 - 439	8,97 - 447	8,97 - 453	12,2 - 440	11,5 - 380	11 - 453	8,79 - 421	13,8 - 471
38	12,9 - 435	9,36 - 423	8,52 - 462	6,73 - 460	11,6 - 496	12,2 - 465	12,9 - 495	7,12 - 443	6,12 - 396	12 - 457
39	12,7 - 443	10,9 - 418	10,7 - 429	12,2 - 555	12,9 - 435	11 - 474	11,9 - 483	10,7 - 459	6,28 - 389	8,67 - 425
40	11,5 - 459	10,7 - 429	10,5 - 450	13,9 - 486	7,31 - 424	12 - 477	9,57 - 464	9,94 - 405	6,28 - 389	12,8 - 460
41	8,91 - 447	8,52 - 462	9,76 - 449	8,92 - 438	11,8 - 465	11 - 474	11 - 474	9,52 - 404	12,2 - 439	10,7 - 459
42	11,2 - 451	9,76 - 449	9,76 - 448	10 - 455	12,2 - 465	9,81 - 430	12,7 - 500	8,92 - 438	8,66 - 388	12,5 - 453
43	7,31 - 424	14,8 - 498	10,7 - 498	8,97 - 447	14 - 469	10,5 - 424	12,4 - 462	6,62 - 429	10,1 - 497	12,5 - 461
44	10,1 - 437	10,5 - 462	13,6 - 475	11 - 453	11,8 - 475	8,88 - 446	11,2 - 499	12,7 - 460	15,2 - 485	9,71 - 428
45	9,65 - 496	13,4 - 529	11,5 - 438	6,62 - 429	9,57 - 429	12,6 - 466	12 - 477	12,5 - 461	9,59 - 443	15 - 524
46	11,2 - 459	11,6 - 496	9,21 - 453	13 - 507	11,8 - 453	13,5 - 477	9,99 - 388	11,4 - 458	12,6 - 433	10,2 - 448
47	15 - 488	8,68 - 442	10,6 - 444	9,99 - 388	11,6 - 437	11,2 - 453	11 - 474	12 - 457	5,87 - 462	15,2 - 523
48	13,5 - 477	10,7 - 429	9,2 - 499	9,57 - 464	13,1 - 460	15 - 488	8,88 - 446	10,8 - 455	12,6 - 433	13,5 - 471
49	12,2 - 440	8,52 - 462	14,7 - 484	11,9 - 483	10,2 - 463	11,3 - 460	10,5 - 424	11,6 - 434	10 - 493	11,4 - 458
50	11,2 - 453	9,76 - 449	11,9 - 474	11,5 - 380	10,1 - 437	12,7 - 443	9,81 - 430	8,38 - 433	6,12 - 396	16,6 - 497
51	7,91 - 439	14,8 - 498	11 - 465	12,4 - 462	7,95 - 442	8,91 - 447	13,9 - 486	9,33 - 456	9,16 - 437	9,67 - 410
52	9,57 - 429	11,9 - 474	11,1 - 450	13,9 - 486	14 - 469	11,2 - 451	12,2 - 555	10,4 - 469	10,1 - 497	12,7 - 460
53	11,8 - 475	11,5 - 438	11,5 - 434	12,1 - 467	7,91 - 439	8,97 - 453	6,48 - 400	15 - 524	9,26 - 453	12,5 - 461
54	7,95 - 442	11 - 465	9,63 - 461	12,7 - 500	12,2 - 440	11,8 - 465	12,1 - 467	15,2 - 523	8,79 - 421	12,5 - 453

55	10,2 - 463	10,3 - 488	9,58 - 456	6,48 - 400	11,2 - 453	12,9 - 435	13 - 507	8,67 - 425	9,59 - 443	8,67 - 425
56	11,6 - 437	10,9 - 418	9,56 - 448	10,5 - 424	13,5 - 477	9,65 - 496	11,9 - 483	12,5 - 453	6,21 - 441	10,2 - 448
57	13,1 - 460	10,7 - 498	9,2 - 499	8,88 - 446	12,6 - 466	14 - 469	11,5 - 380	9,32 - 450	16,1 - 514	15,2 - 523
58	10,7 - 447	9,27 - 445	11,1 - 450	9,81 - 430	14,4 - 528	16,8 - 496	12 - 477	10,2 - 448	7,97 - 448	7,16 - 400
59	11,3 - 460	9,36 - 423	10,7 - 498	12 - 477	8,19 - 430	14,4 - 528	11 - 474	12,8 - 460	9,59 - 443	9,94 - 405
60	11,8 - 453	13,6 - 475	14,7 - 484	11,2 - 499	10,1 - 437	7,31 - 424	8,88 - 446	8,77 - 433	7,97 - 448	15 - 524
61	15,5 - 543	12,7 - 472	11,9 - 474	5,46 - 409	8,91 - 447	11,8 - 475	10,5 - 424	13,5 - 471	7,4 - 384	9,71 - 428
62	6,21 - 441	9,2 - 499	11,5 - 438	9,85 - 461	11,2 - 459	10,1 - 437	11,9 - 438	16,6 - 497	9,26 - 453	9,52 - 404
63	5,87 - 462	11,5 - 417	11 - 465	10 - 455	7,95 - 442	11,2 - 459	13 - 507	9,71 - 428	10,1 - 497	9,33 - 456
64	14 - 584	13,4 - 529	13,6 - 475	8,97 - 447	11,8 - 465	11,6 - 437	9,85 - 461	10,7 - 459	6,21 - 441	9,67 - 410
65	9,81 - 430	12,7 - 474	8,52 - 462	11,9 - 438	9,65 - 496	13,1 - 460	10 - 455	13,8 - 471	11,5 - 479	13,5 - 471
66	6,28 - 389	9,21 - 453	9,76 - 449	11,9 - 438	15 - 488	9,57 - 429	9,81 - 430	9,52 - 404	11,5 - 479	16,6 - 497
67	11,6 - 434	14,7 - 484	9,76 - 448	9,85 - 461	8,91 - 447	11,5 - 459	13,4 - 487	9,67 - 410	6,28 - 389	12,8 - 460
68	12,2 - 461	14,6 - 445	12,5 - 455	15,8 - 507	9,93 - 433	12,2 - 466	10,2 - 440	10,8 - 406	9,71 - 428	12,6 - 530
A	13,6 - 452	12,3 - 466	15,2 - 479	10,9 - 466	13,6 - 463	11 - 426	11,5 - 462	15,2 - 497	10,3 - 541	12,4 - 436
B	9,76 - 454	10,8 - 470	10,9 - 401	10,8 - 404	13,4 - 460	12 - 436	9,33 - 381	12,1 - 431	15,6 - 556	13,1 - 473
C	11,9 - 452	13 - 453	9,73 - 420	13,6 - 481	11,2 - 465	13,5 - 490	10,6 - 421	10,1 - 411	9,84 - 450	12,1 - 507
D	9,72 - 411	15,8 - 552	10,7 - 469	10,5 - 415	12,8 - 496	12,8 - 473	16,7 - 549	13,4 - 489	14,5 - 491	9,75 - 408
E	13,2 - 475	14,5 - 487	11,1 - 469	12,3 - 504	12,4 - 494	13,8 - 463	11,6 - 441	9,92 - 404	12 - 445	9,33 - 415
F	8,73 - 387	11 - 471	14,4 - 475	10,7 - 397	9,18 - 386	13,3 - 459	11,2 - 422	11,6 - 450	8,56 - 392	14,1 - 483
G	10,4 - 442	13,6 - 449	12,8 - 454	11 - 456	13,1 - 482	15,4 - 478	14,7 - 512	12,7 - 461	13,3 - 445	16,5 - 521
H	12,3 - 460	14,5 - 513	11,5 - 419	10,9 - 448	11,2 - 449	12,7 - 461	8,96 - 387	9,96 - 439	13,9 - 463	9,94 - 406
I	15,7 - 526	13 - 433	9,65 - 449	10,1 - 441	15,9 - 530	13,1 - 474	11,1 - 424	12,1 - 464	12,7 - 488	12,4 - 475
J	9,53 - 422	10,3 - 421	9,44 - 412	9,46 - 384	11,4 - 472	17,5 - 530	13,7 - 492	14,5 - 495	11,8 - 463	12,8 - 476
K	8,13 - 413	8,18 - 418	11,9 - 448	11,6 - 467	12,8 - 482	8,01 - 414	12,3 - 471	9,42 - 426	11,5 - 467	8,41 - 444
L	12,3 - 467	14,6 - 495	9,19 - 409	13,4 - 471	13,4 - 509	11,6 - 508	13 - 474	12,9 - 457	12,2 - 434	8,53 - 382

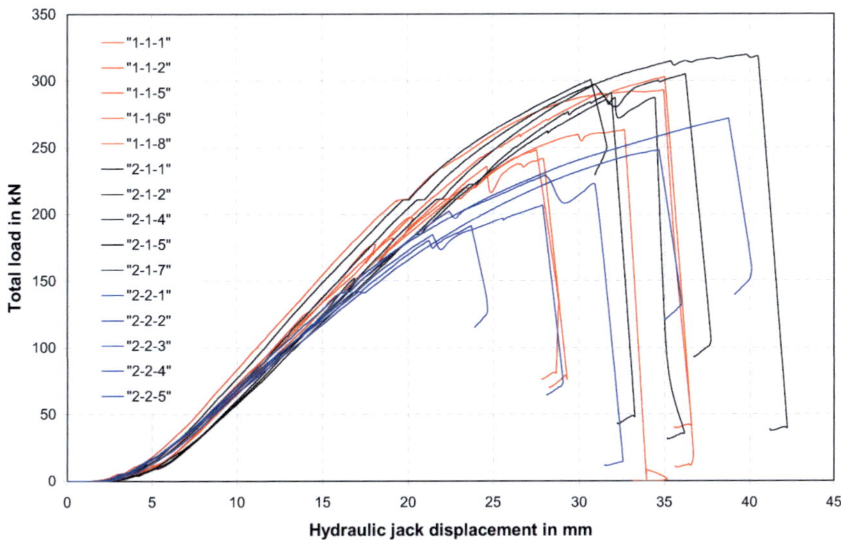

Fig. 67   Load-displacement behaviour for short beams

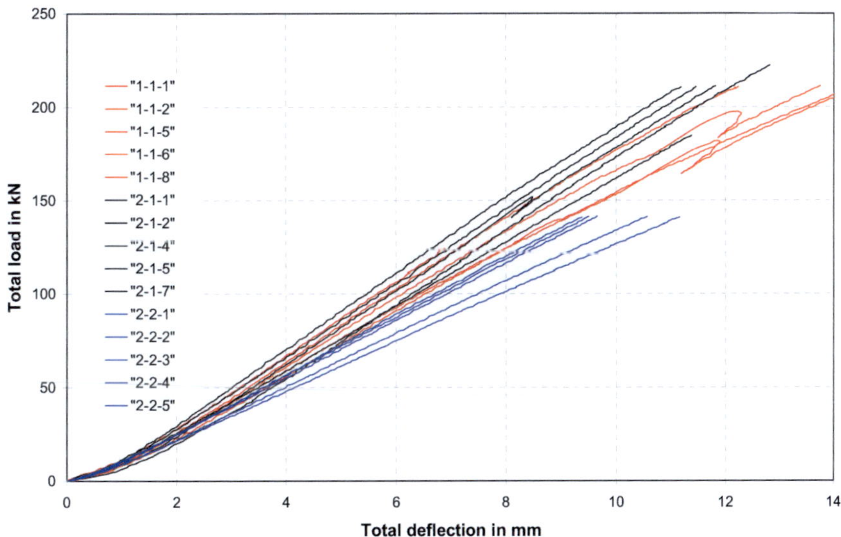

Fig. 68   Load-displacement behaviour for short beams

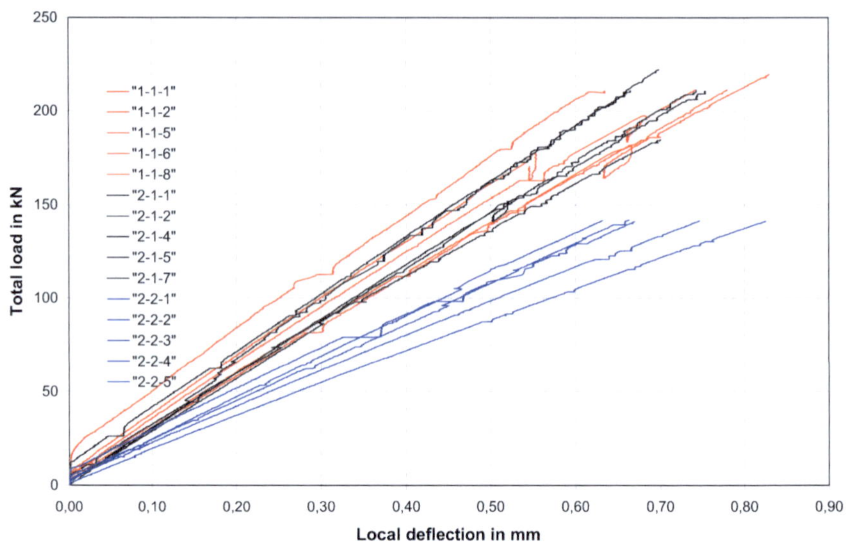

Fig. 69   Load-displacement behaviour for short beams

Fig. 70   Typical failure in bending for CLT beam (left). Failure at the CLT beam end (right)

Fig. 71   Typical failure for DLT beams with five plies

Fig. 72   Test set-up and typical failure for DLT beams with four plies

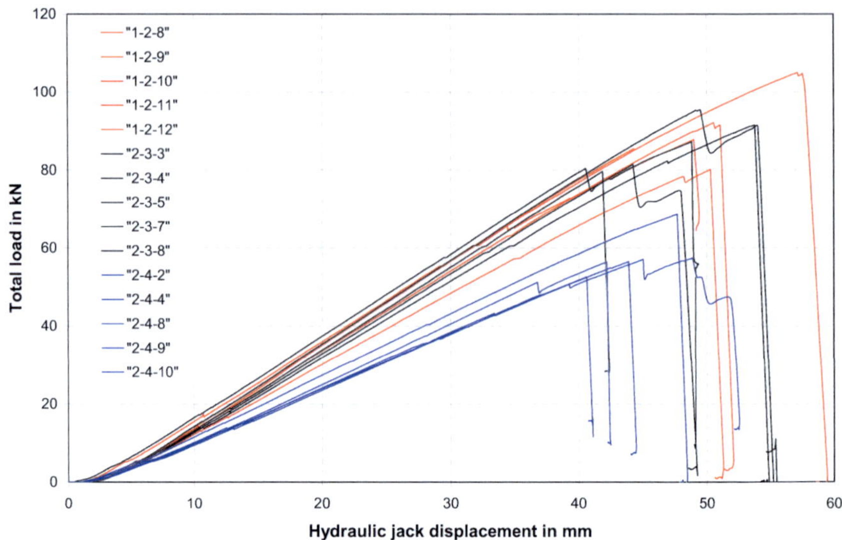

Fig. 73   Load-displacement behaviour for long beams

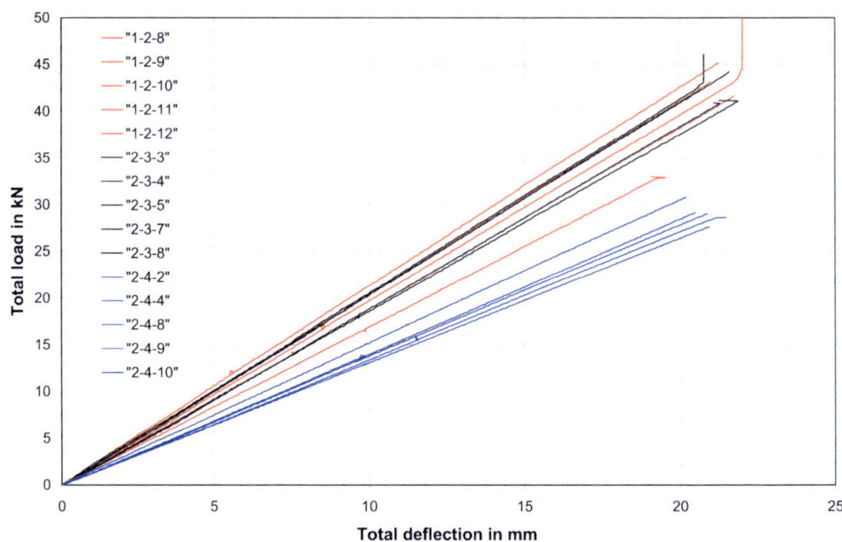

Fig. 74   Load-displacement behaviour for long beams

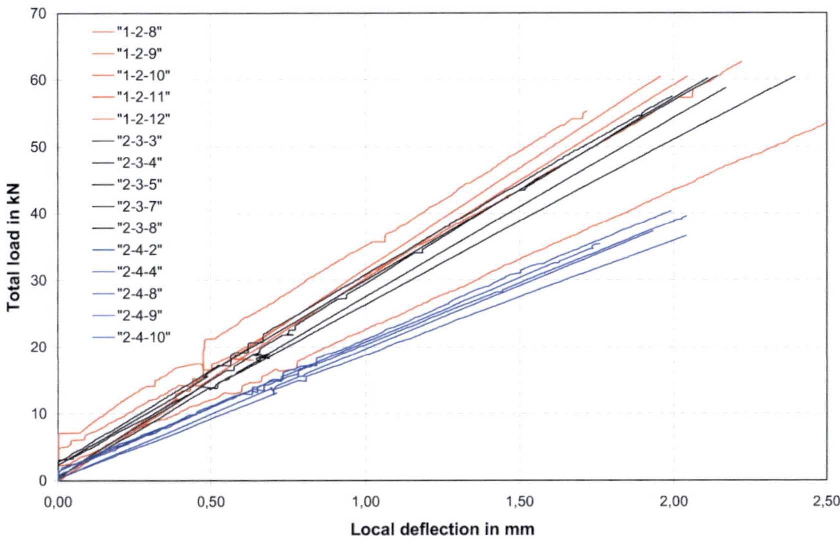

Fig. 75   Load-displacement behaviour for long beams

Fig. 76   Testing assembly for long beams (left). Loaded long beam before bending failure (right)

Fig. 77   Typical failure for long beams displayed as an example on a DLT beam with five plies

Fig. 78   DLT beam with four plies before and after bending failure

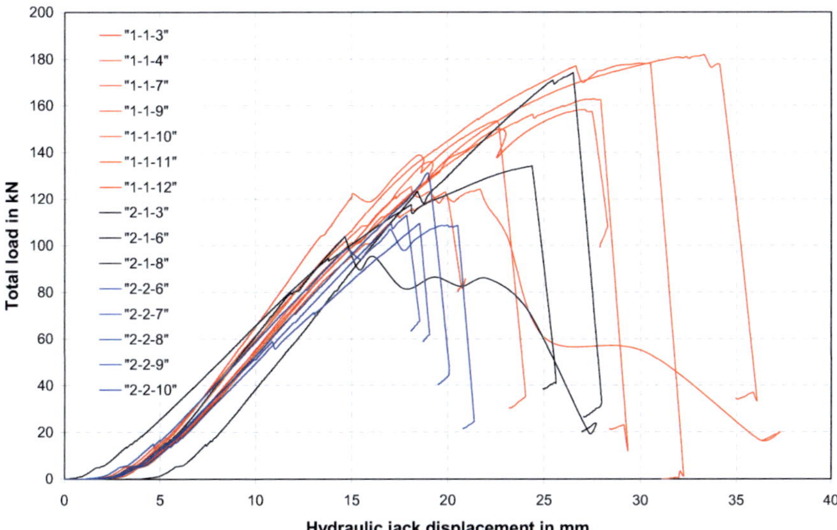

Fig. 79   Load-displacement behaviour for short beams with notches

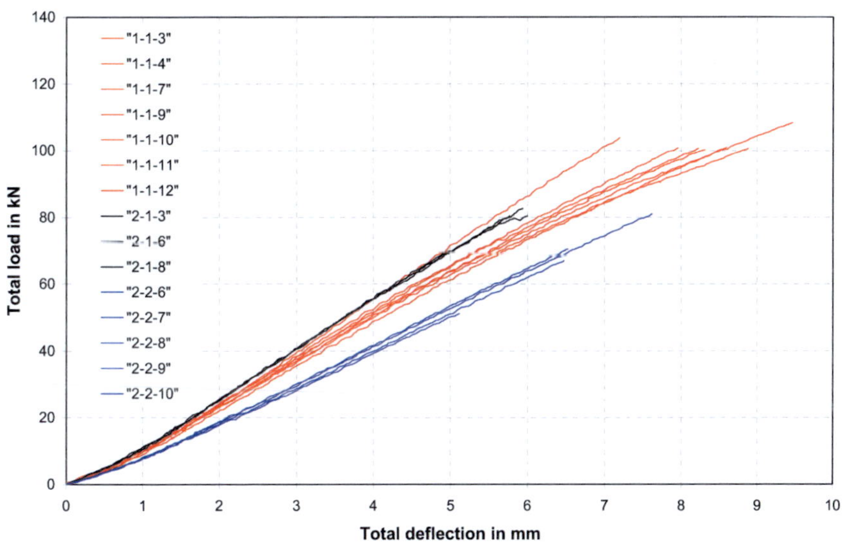

Fig. 80   Load-displacement behaviour for short beams with notches

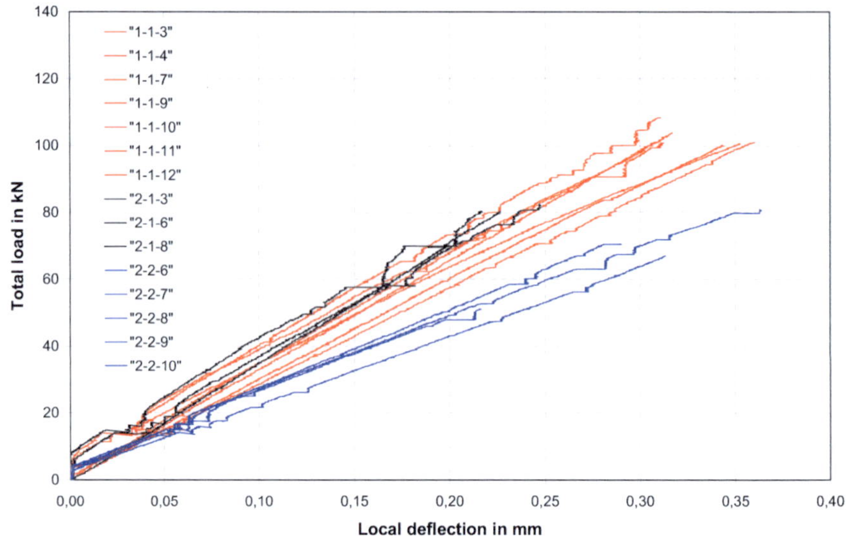

Fig. 81   Load-displacement behaviour for short beams with notches

Fig. 82   Typical failure of the reinforcing layer in short CLT beams with notches

Fig. 83   Failure of the reinforcing layer in short DLT beams with five plies and with notches

Fig. 84   Testing assembly for beams with notches (left). Failure of the reinforcing layer in short DLT beams with four plies and with notches (right)

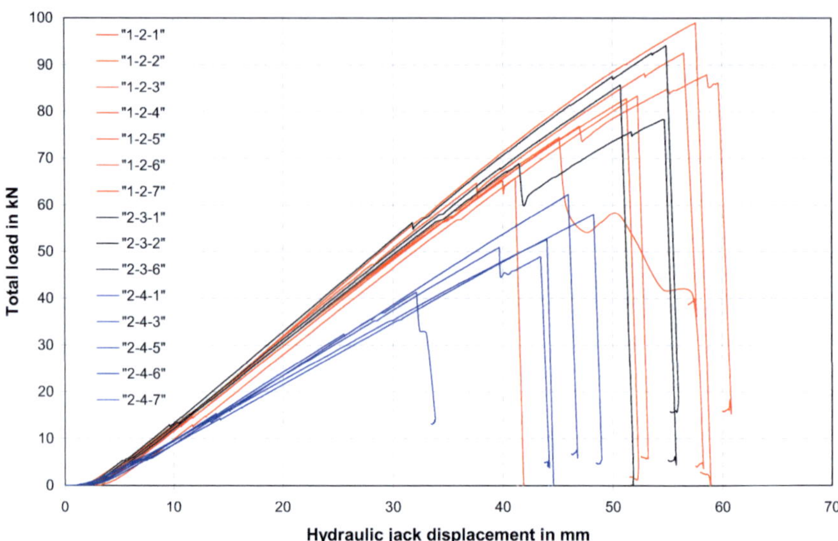

Fig. 85   Load-displacement behaviour for long beams with holes

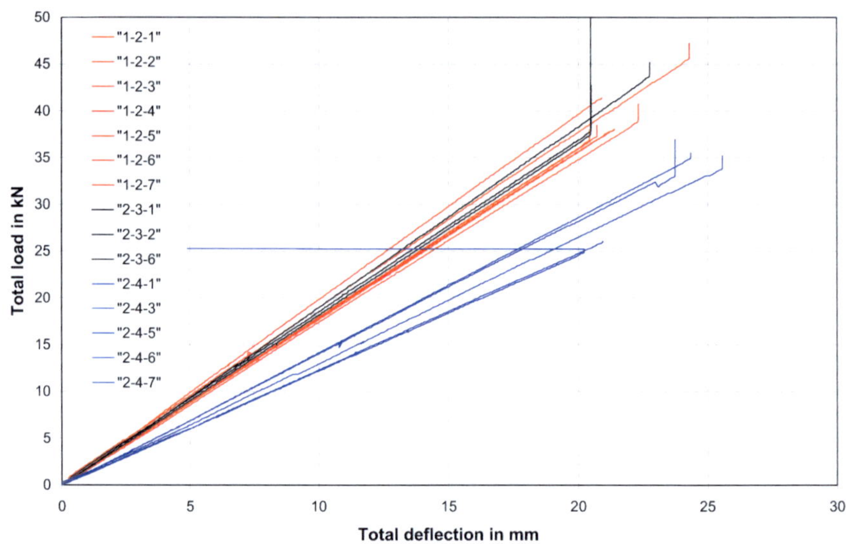

Fig. 86   Load-displacement behaviour for long beams with holes

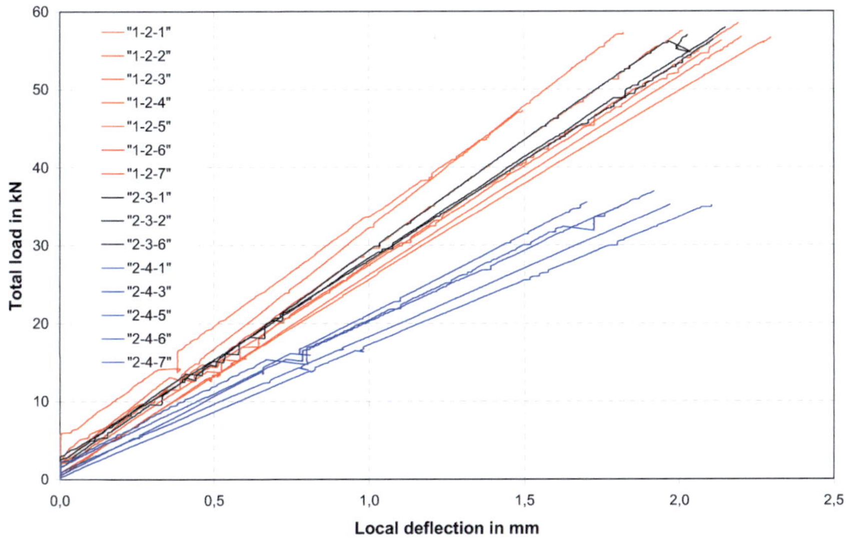

Fig. 87   Load-displacement behaviour for long beams with holes

Fig. 88   Testing assembly for beams with holes before testing (left) and before failure (right)

Fig. 89   Bending failure for CLT beams (left) and for DLT beams (right)

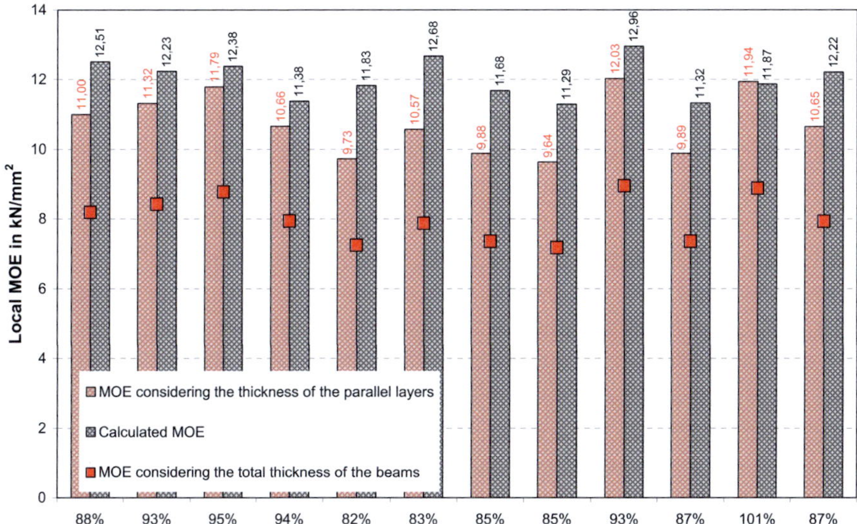

Fig. 90 $E_{0,net}$ (E_0) versus E_{cal} for specimens 1-1 (short CLT with five plies)

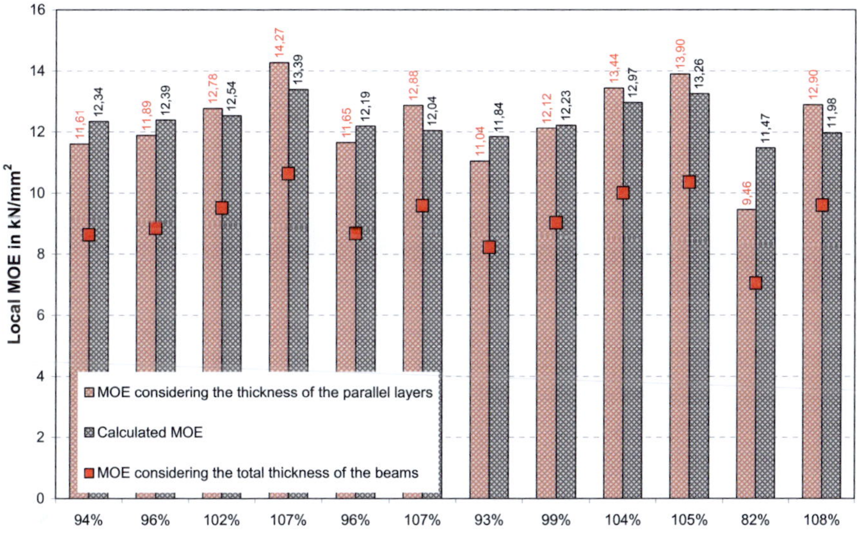

Fig. 91 $E_{0,net}$ (E_0) versus E_{cal} for specimens 1-2 (long CLT with five plies)

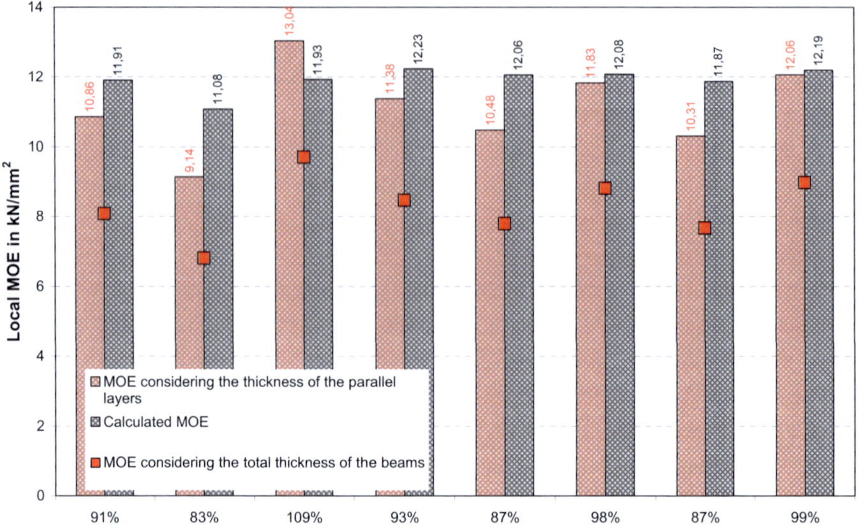

Fig. 92 $E_{0,net}$ (E_0) versus E_{cal} for specimens 2-1 (short DLT with five plies)

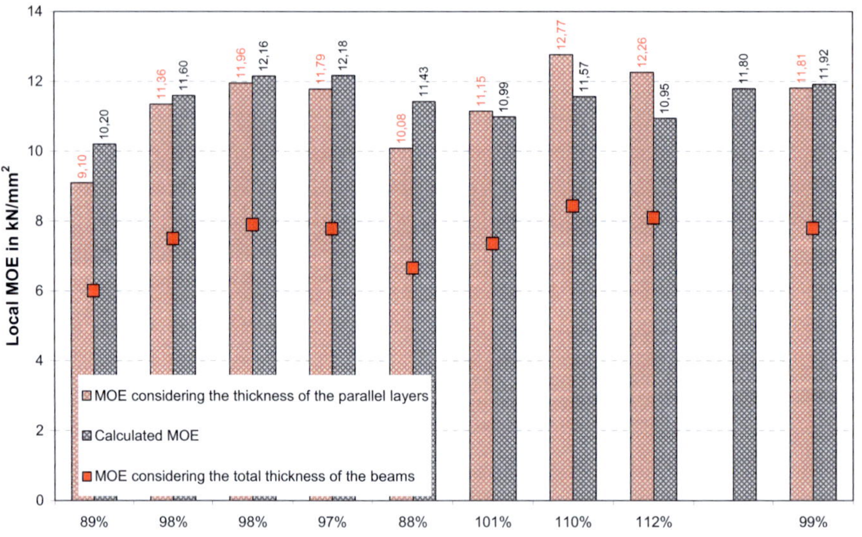

Fig. 93 $E_{0,net}$ (E_0) versus E_{cal} for specimens 2-2 (short DLT with four plies)

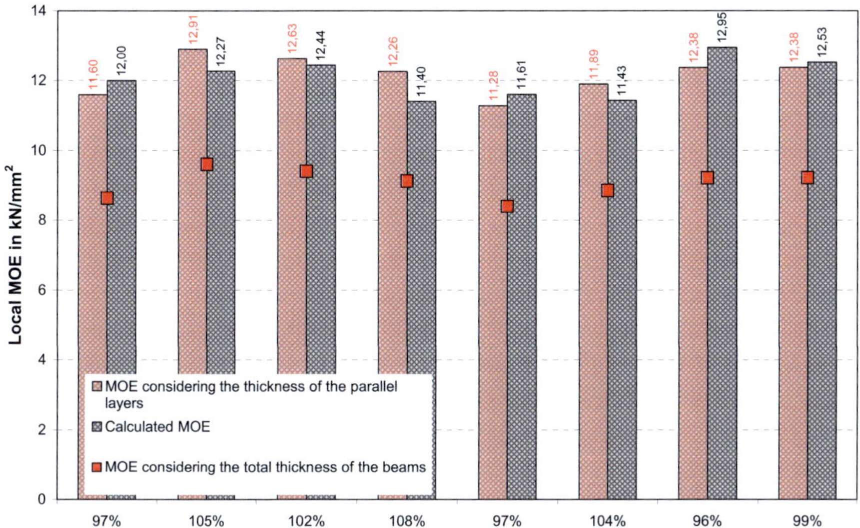

Fig. 94 $E_{0,net}$ (E_0) versus E_{cal} for specimens 2-3 (long DLT with five plies)

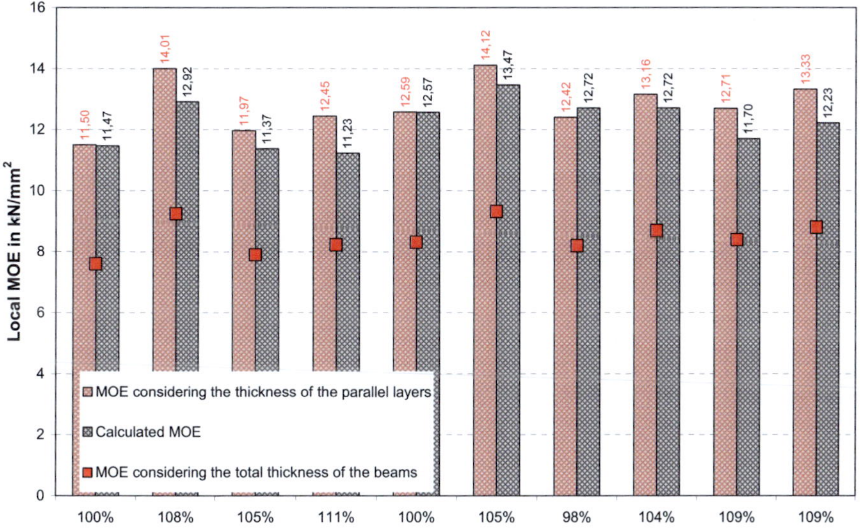

Fig. 95 $E_{0,net}$ (E_0) versus E_{cal} for specimens 2-4 (long DLT with four plies)

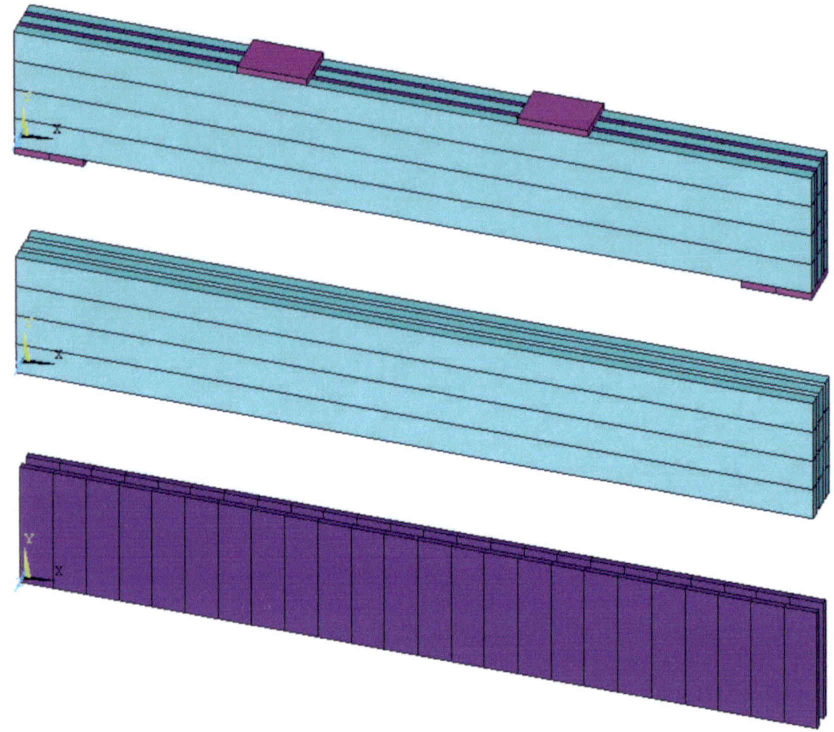

Fig. 96　　FE system for specimen 1-1 (short CLT with five plies)

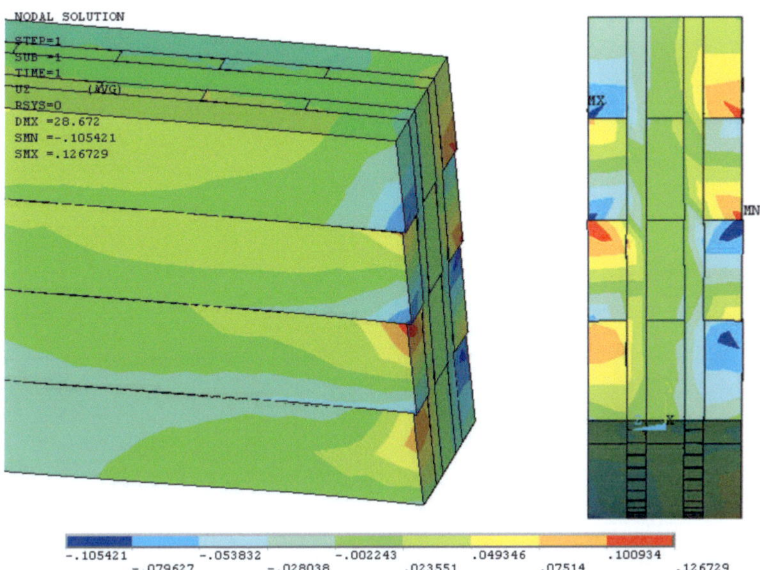

Fig. 97   Beam end with deflection in z-direction (Sp.1-1, five times distorted)

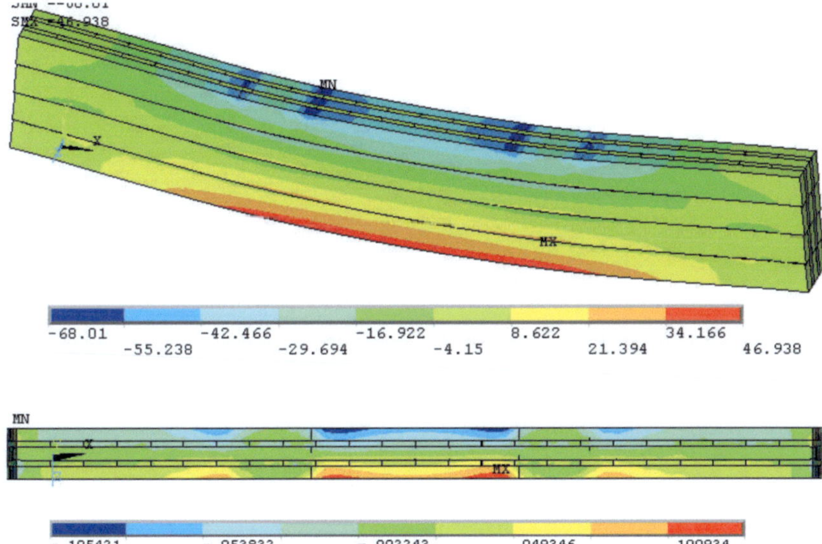

Fig. 98   Bending stresses (top) and deflection in z-direction (bottom) (five times distorted)

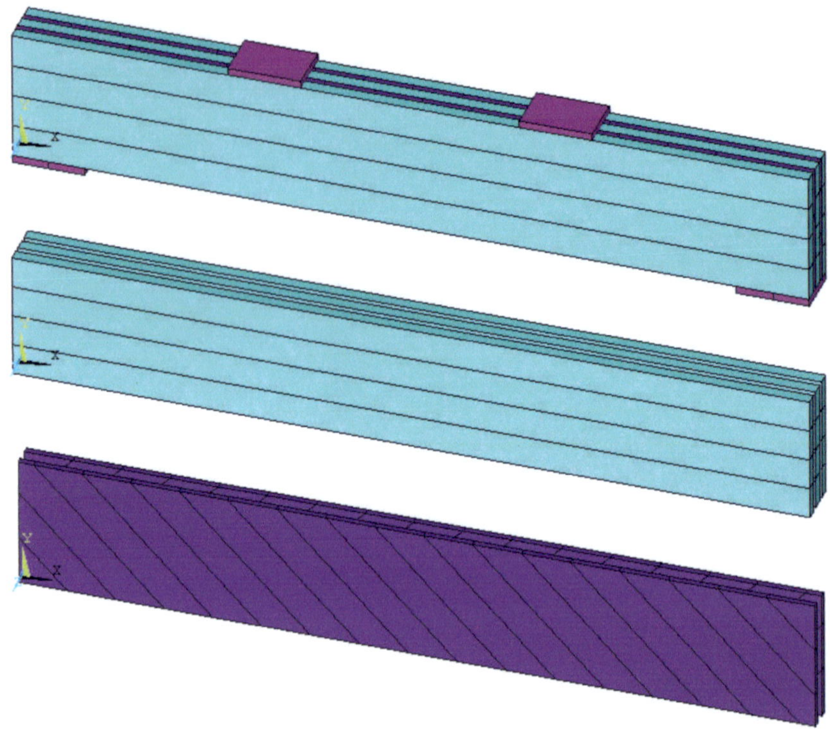

Fig. 99   FE system for specimen 2-1 (short DLT with five plies)

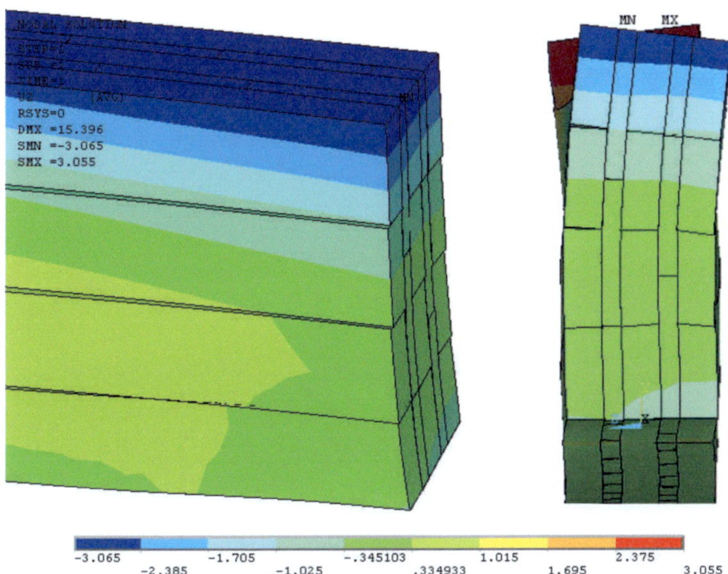

Fig. 100 Beam end with deflection in z-direction (Sp.2-1, five times distorted)

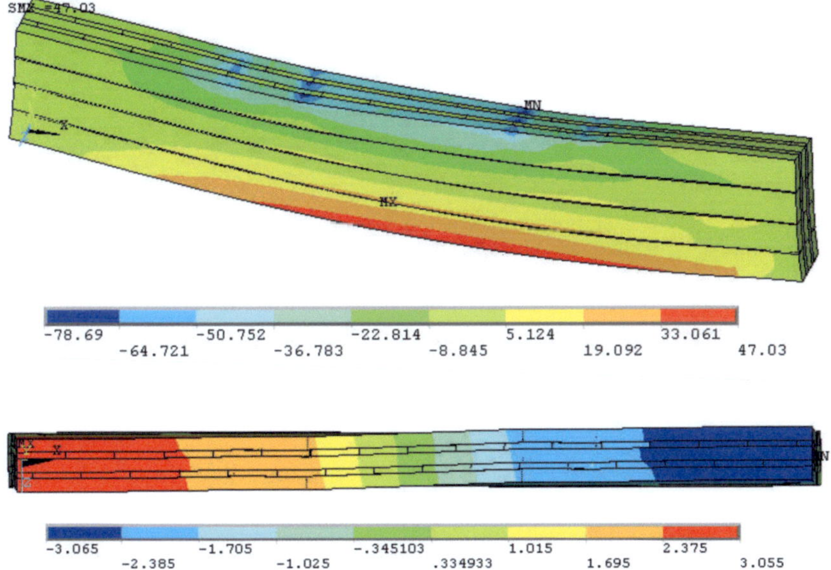

Fig. 101 Bending stresses (top) and deflection in z-direction (bottom) (five times distorted)

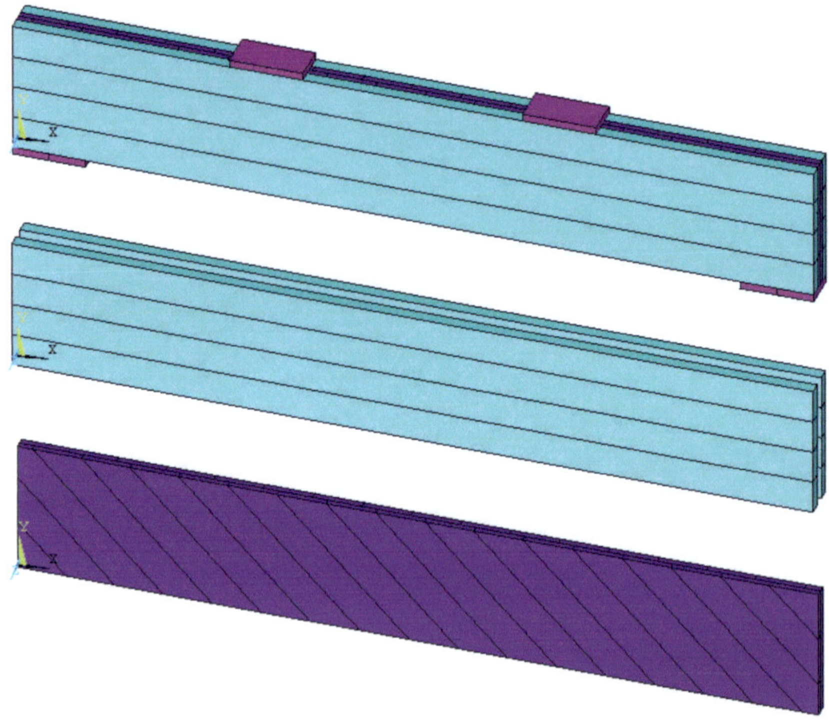

Fig. 102   FE system for specimen 2-2 (short DLT with four plies)

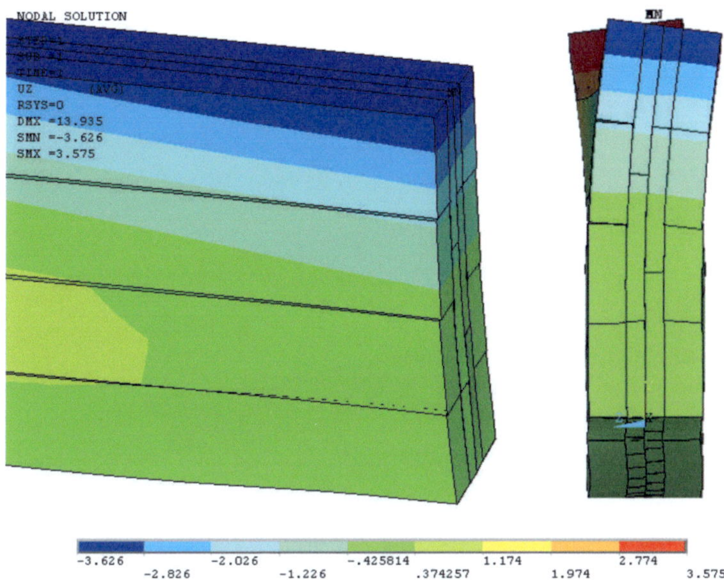

Fig. 103   Beam end with deflection in z-direction (Sp.2-2, five times distorted)

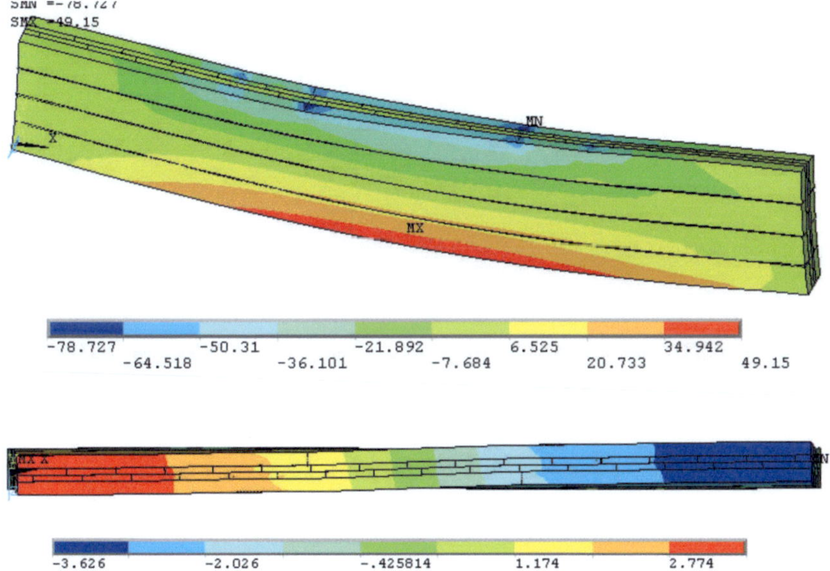

Fig. 104   Bending stresses (top) and deflection in z-direction (bottom) (five times distorted)

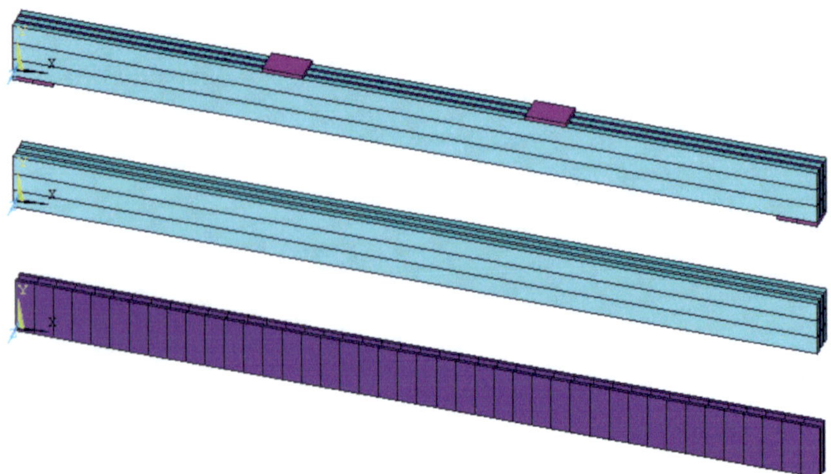

Fig. 105   FE system for specimen 1-2 (long CLT with five plies)

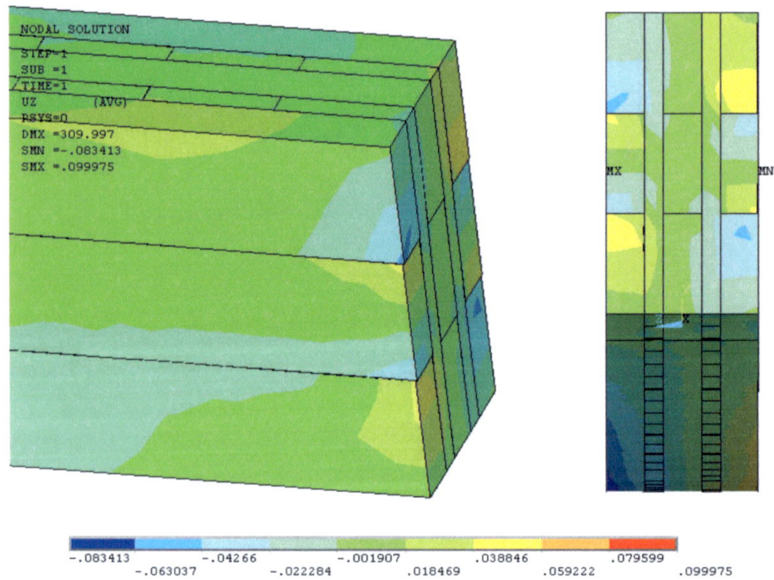

Fig. 106   Beam end with deflection in z-direction (Sp.1-2, three times distorted)

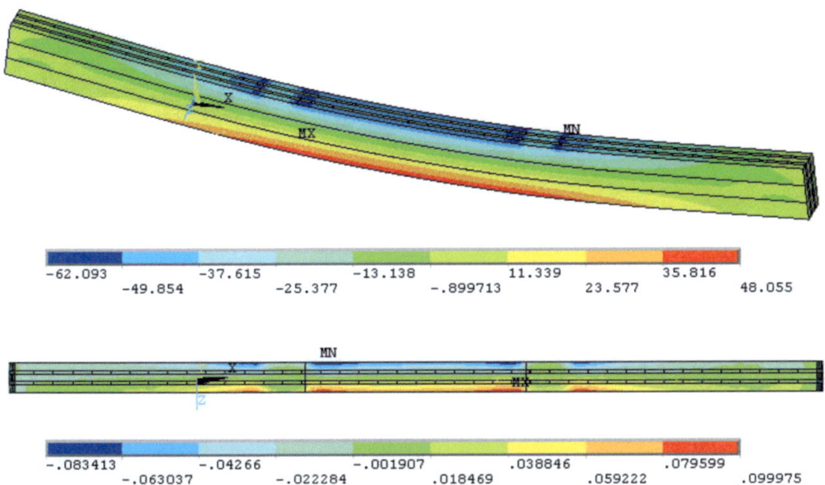

Fig. 107 Bending stresses (top) and deflection in z-direction (bottom) (three times distorted)

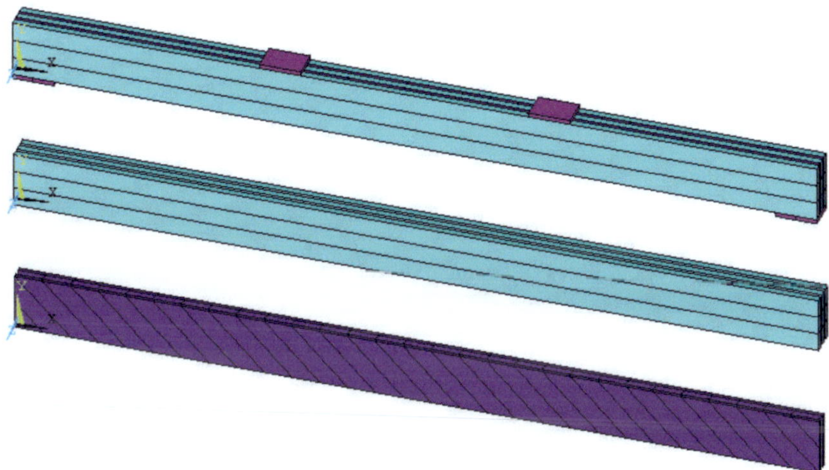

Fig. 108 FE system for specimen 2-3 (long DLT with five plies)

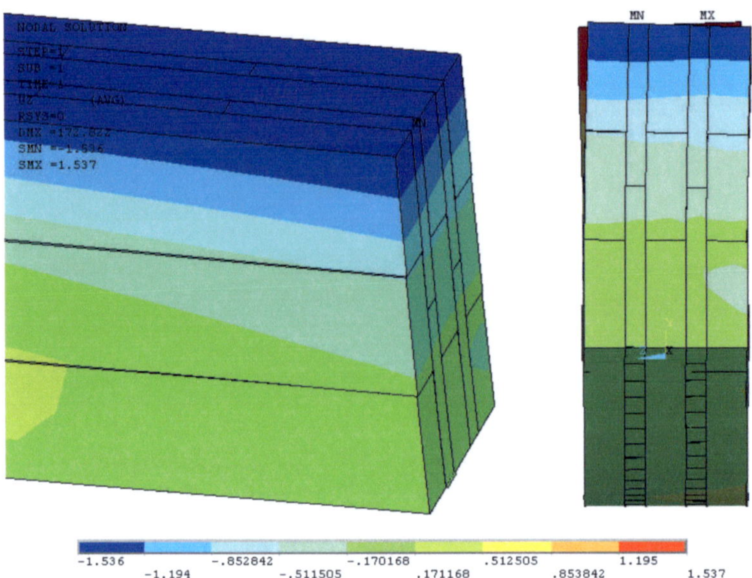

Fig. 109   Beam end with deflection in z-direction (Sp.2-3, three times distorted)

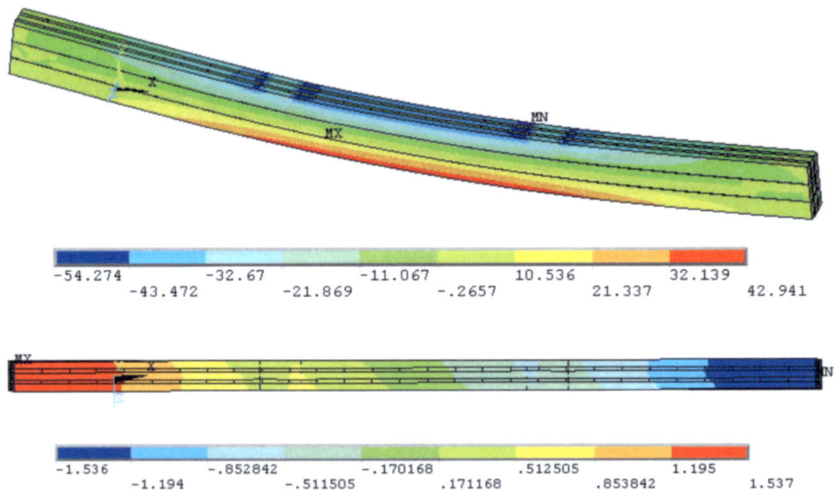

Fig. 110   Bending stresses (top) and deflection in z-direction (bottom) (three times
           distorted)

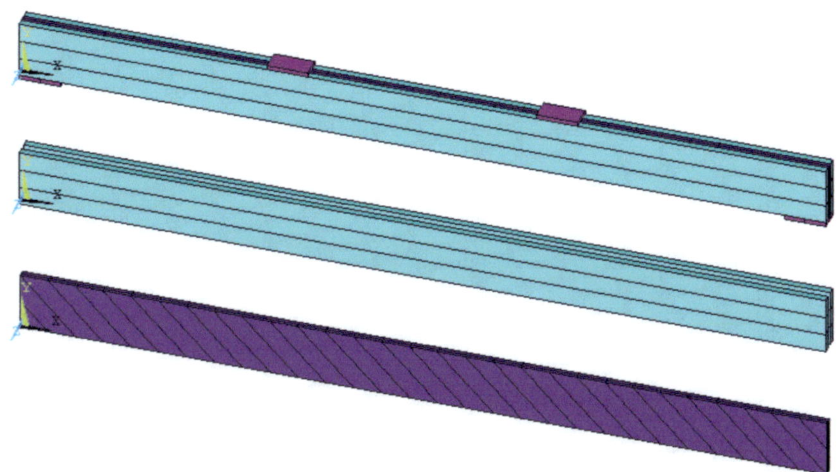

Fig. 111   FE system for specimen 2-4 (long DLT with four plies)

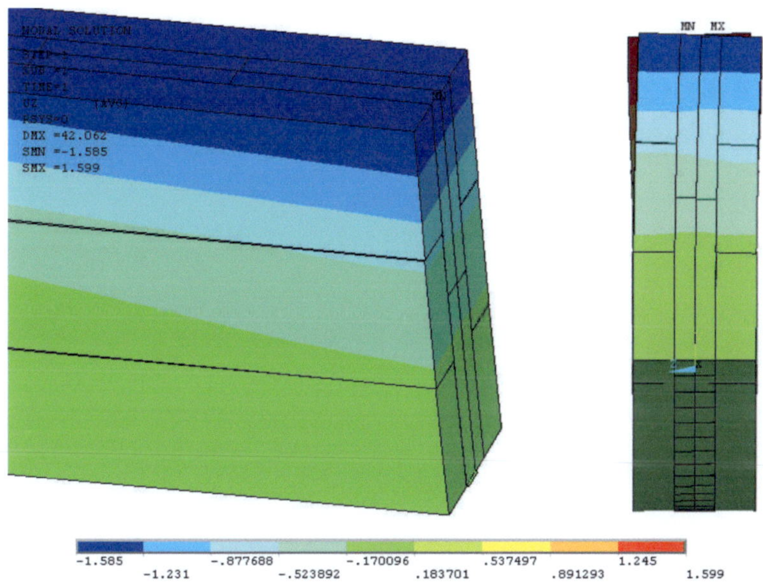

Fig. 112   Beam end with deflection in z-direction (Sp.2-4, three times distorted)

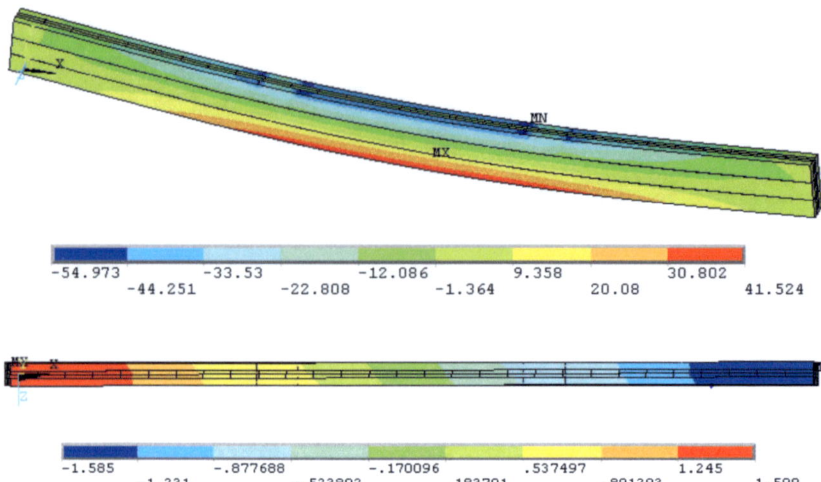

Fig. 113   Bending stresses (top) and deflection in z-direction (bottom) (three times distorted)

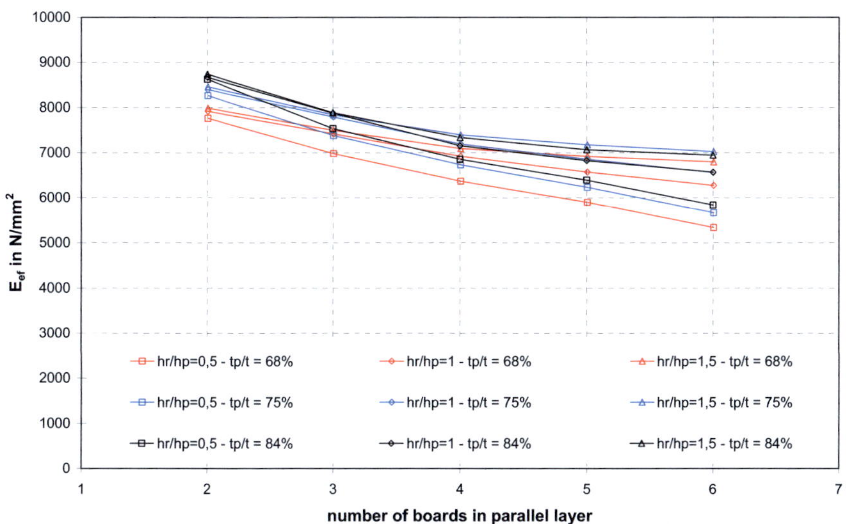

Fig. 114 E_{ef} for short 5-ply CLT beams (Series 1-1)

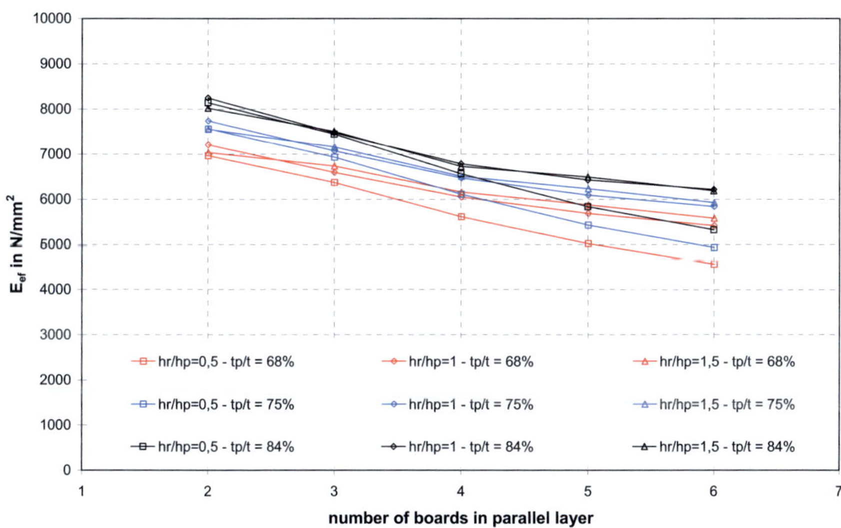

Fig. 115 E_{ef} for short 5-ply DLT beams (Series 2-1)

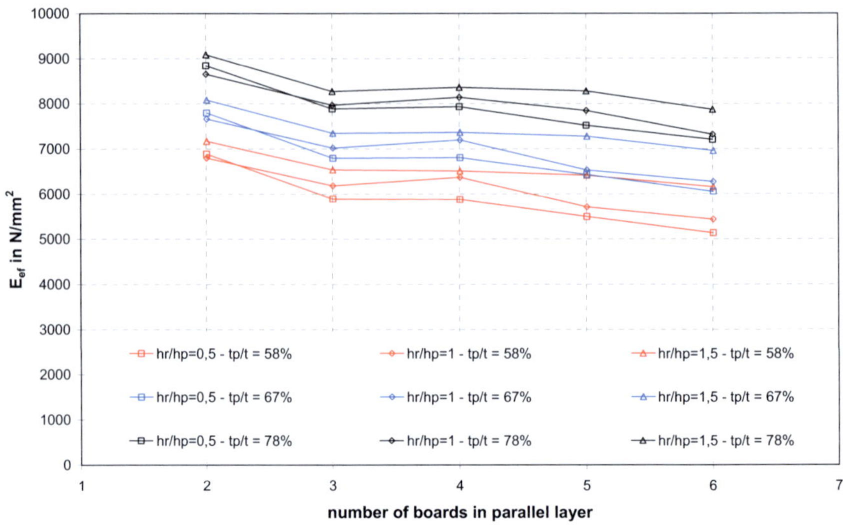

Fig. 116 E_{ef} for short 4-ply DLT beams (Series 2-2)

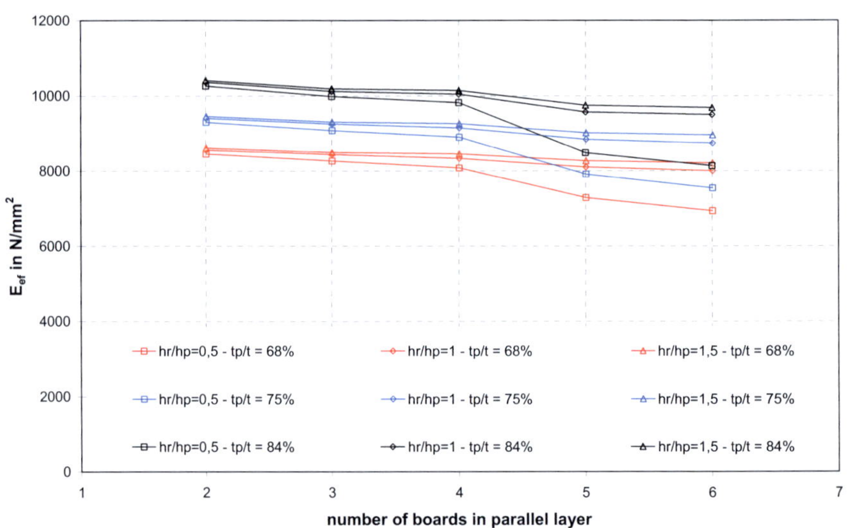

Fig. 117 E_{ef} for long 5-ply CLT beams (Series 1-2)

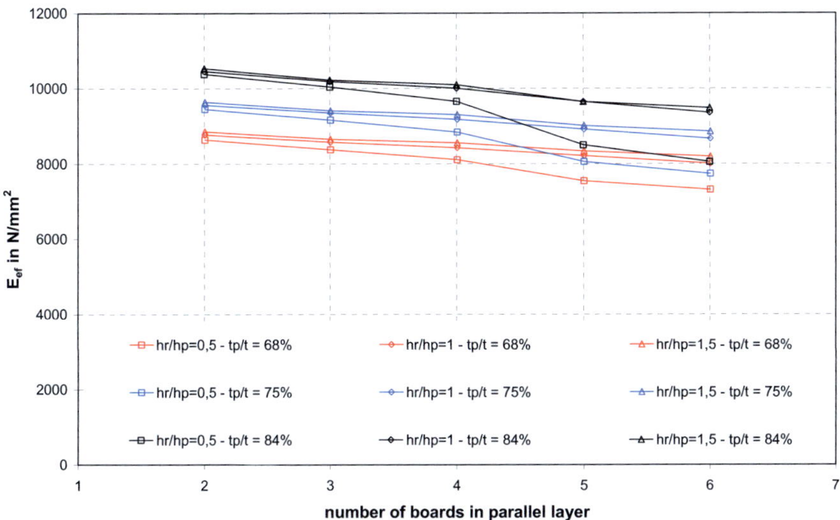

Fig. 118 E_{ef} for long 5-ply DLT beams (Series 2-3)

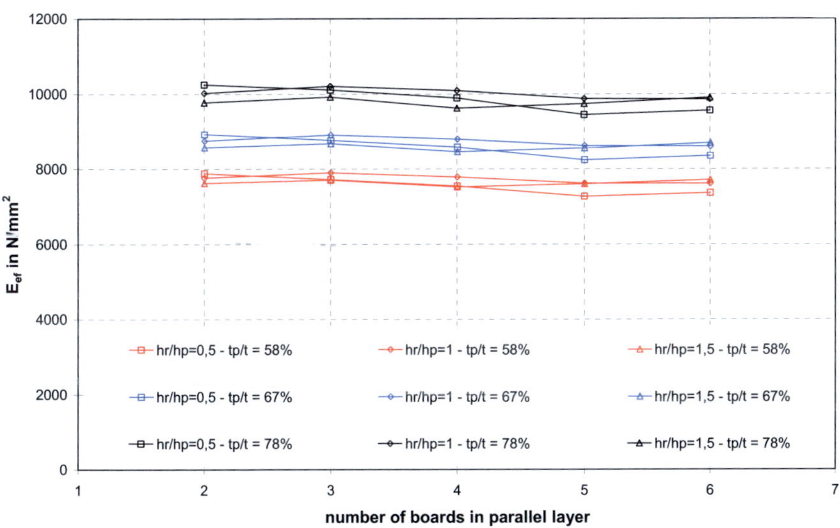

Fig. 119 E_{ef} for long 4-ply DLT beams (Series 2-4)

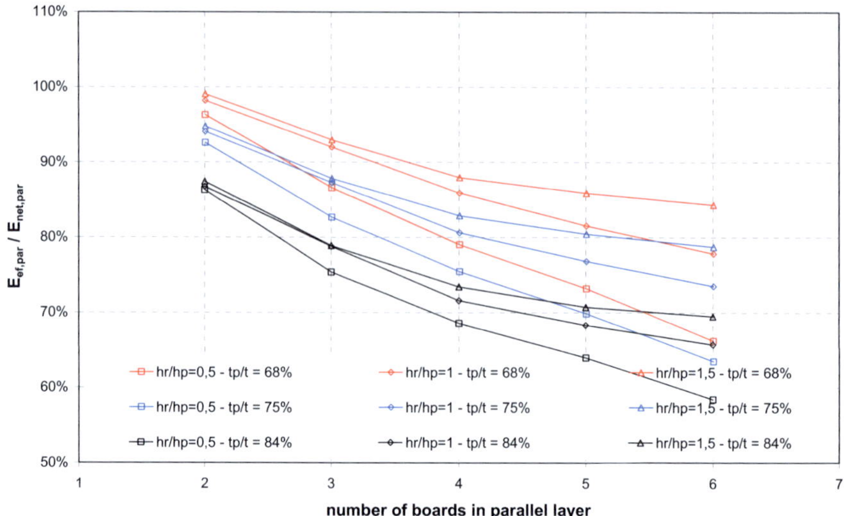

Fig. 120   Ratio between E_{ef} and E_{net} for short 5-ply CLT beams (Series 1-1)

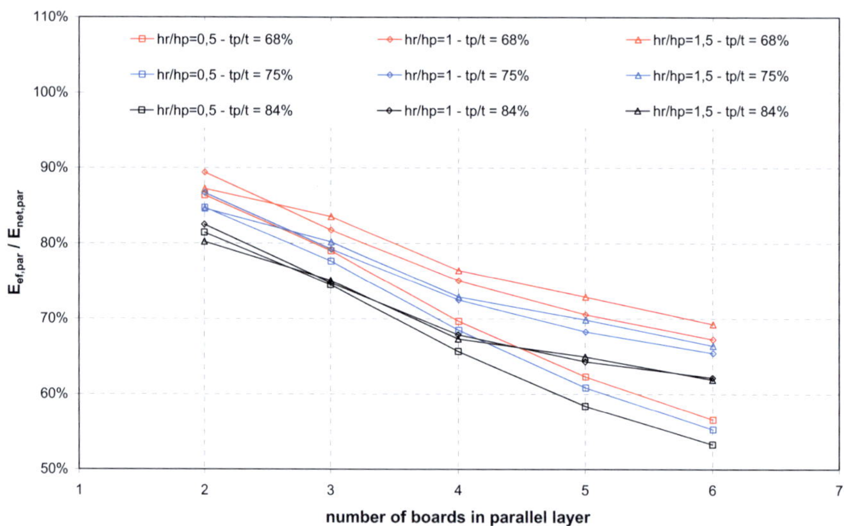

Fig. 121   Ratio between E_{ef} and E_{net} for short 5-ply DLT beams (Series 2-1)

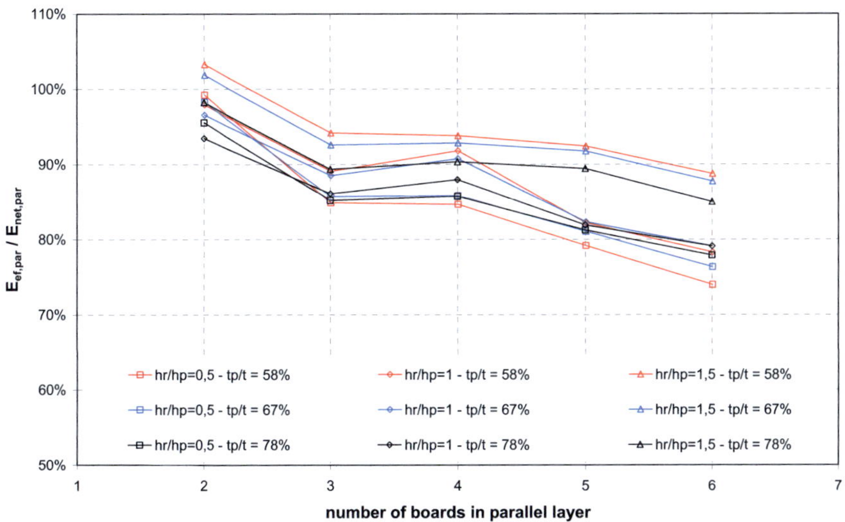

Fig. 122   Ratio between E_{ef} and E_{net} for short 4-ply DLT beams (Series 2-2)

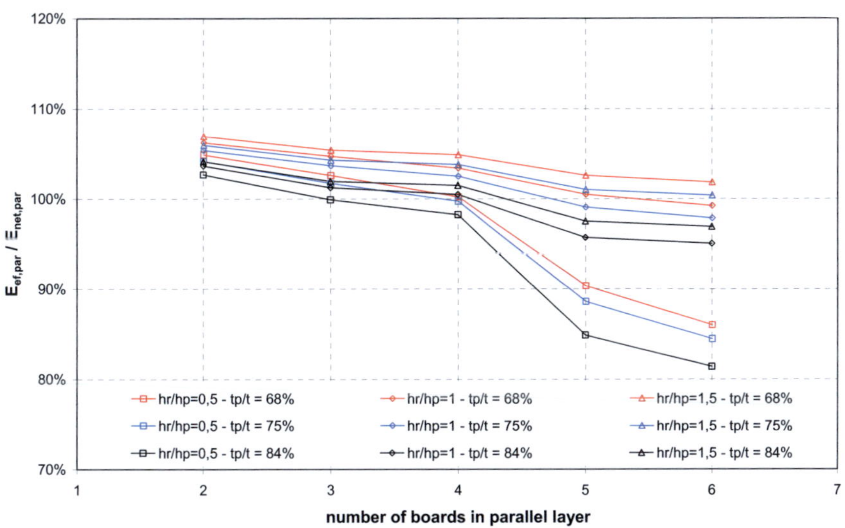

Fig. 123   Ratio between E_{ef} and E_{net} for long 5-ply CLT beams (Series 1-2)

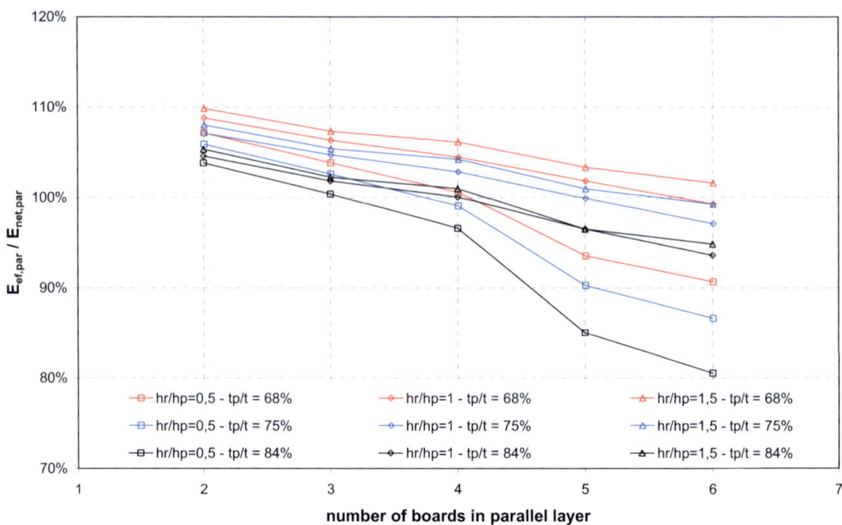

Fig. 124   Ratio between E_{ef} and E_{net} for long 5-ply DLT beams (Series 2-3)

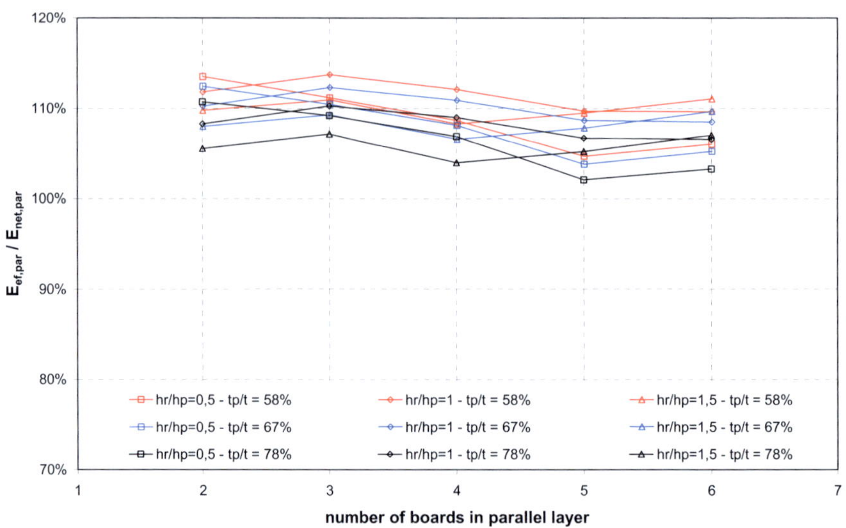

Fig. 125   Ratio between E_{ef} and E_{net} for long 4-ply DLT beams (Series 2-4)

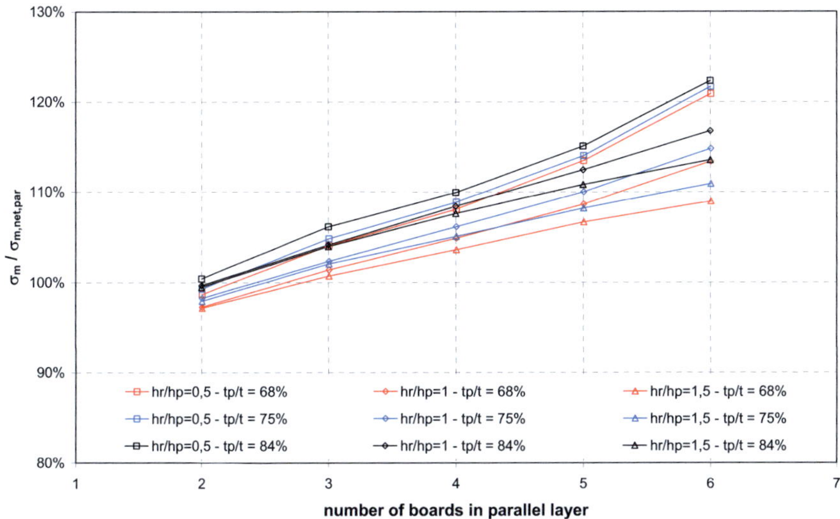

Fig. 126   Ratio between σ_m and $\sigma_{m,net}$ for short 5-ply CLT beams (Series 1-1)

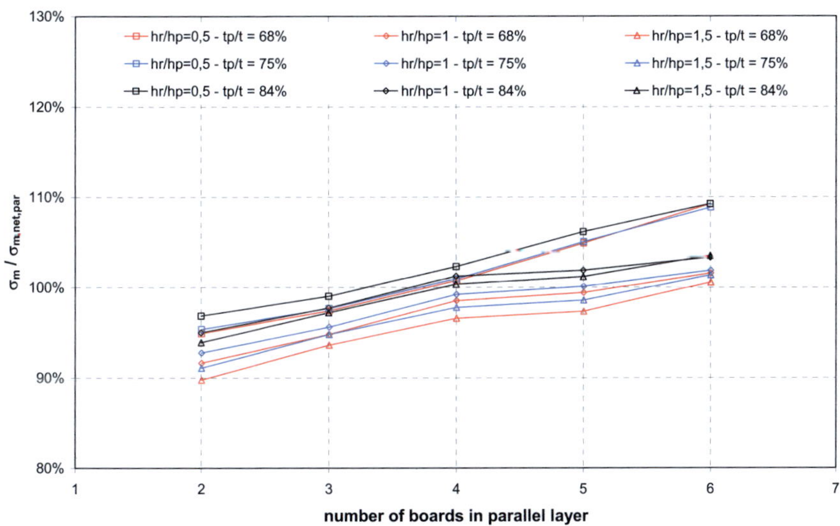

Fig. 127   Ratio between σ_m and $\sigma_{m,net}$ for short 5-ply DLT beams (Series 2-1)

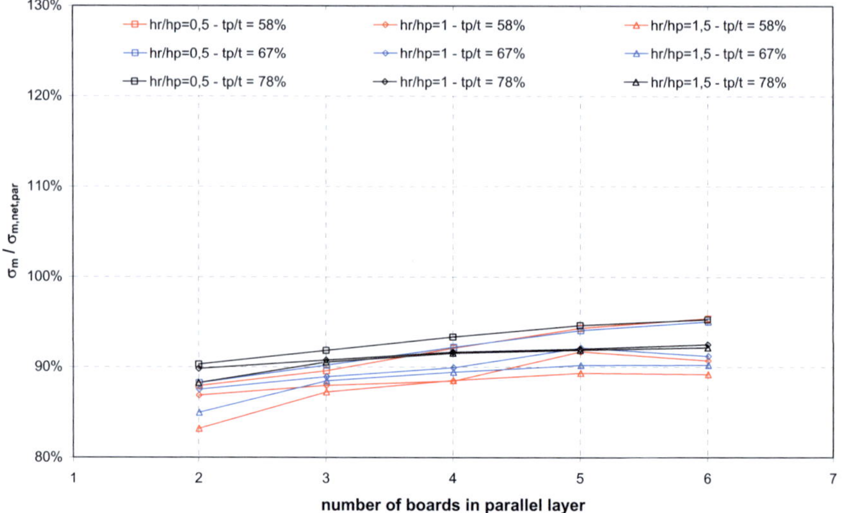

Fig. 128   Ratio between σ_m and $\sigma_{m,net}$ for short 4-ply DLT beams (Series 2-2)

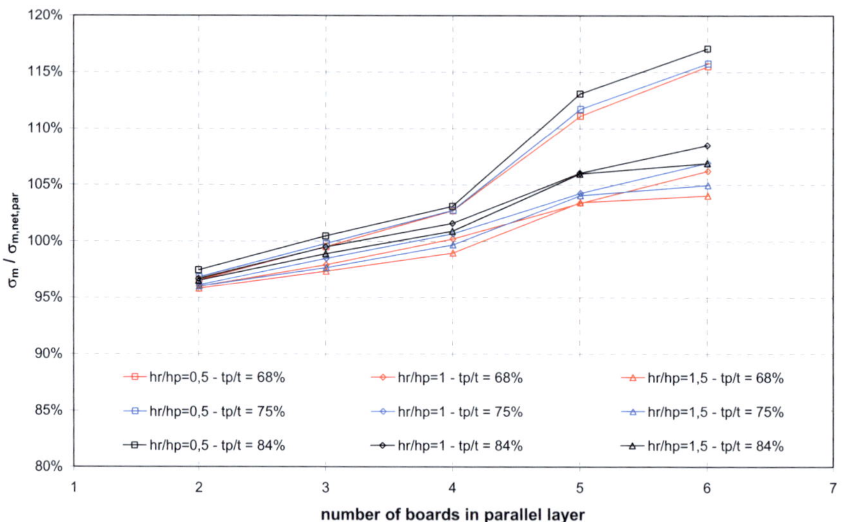

Fig. 129   Ratio between σ_m and $\sigma_{m,net}$ for long 5-ply CLT beams (Series 1-2)

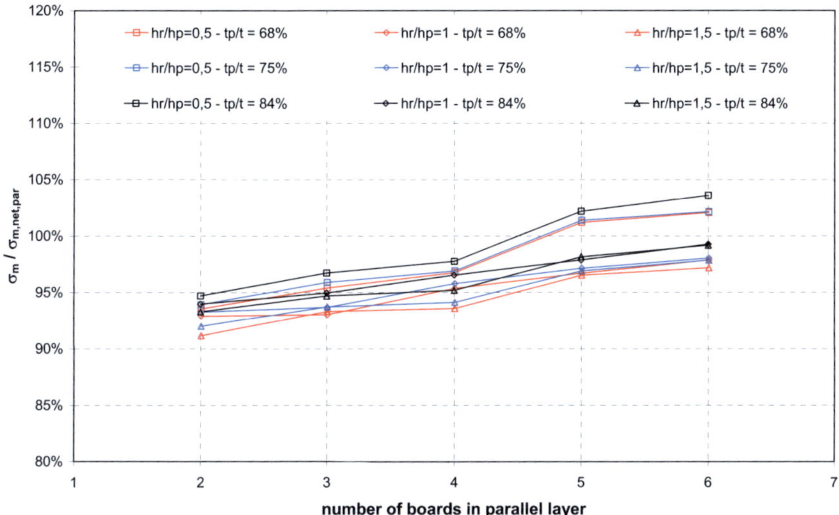

Fig. 130   Ratio between σ_m and $\sigma_{m,net}$ for long 5-ply DLT beams (Series 2-3)

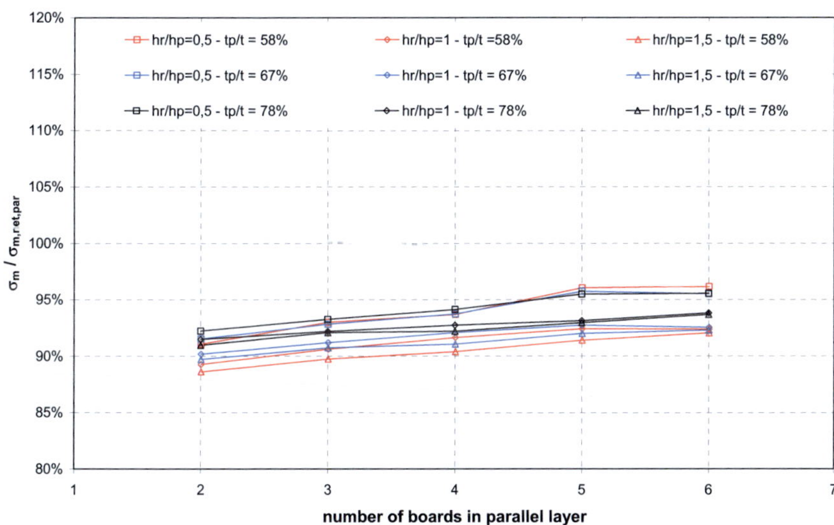

Fig. 131   Ratio between σ_m and $\sigma_{m,net}$ for long 4-ply DLT beams (Series 2-4)

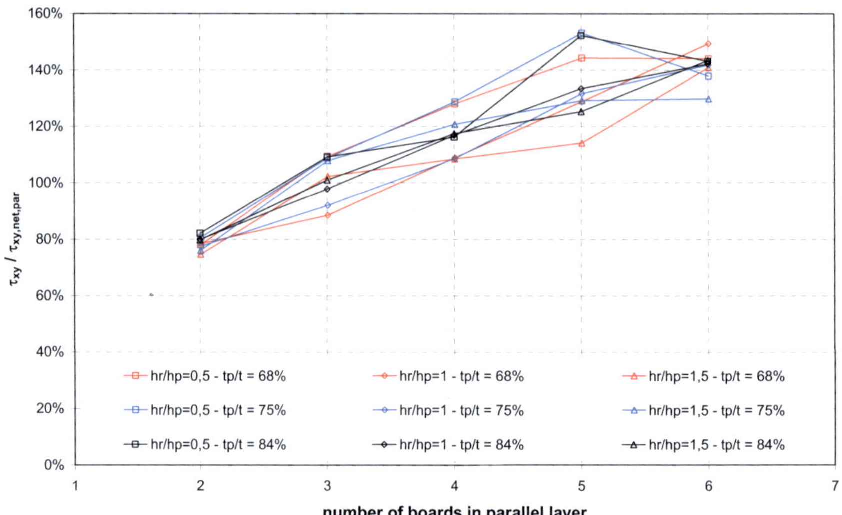

Fig. 132   Ratio between τ_{xy} and $\tau_{xy,net}$ for short 5-ply CLT beams (Series 1-1)

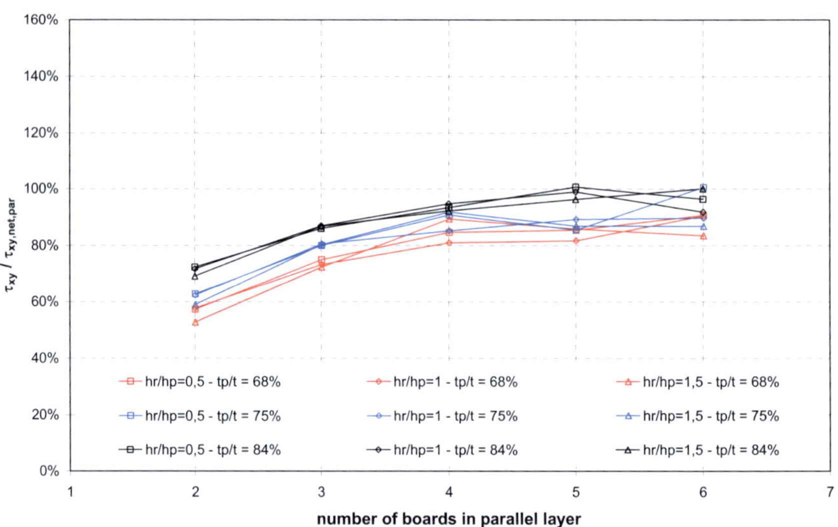

Fig. 133   Ratio between τ_{xy} and $\tau_{xy,net}$ for short 5-ply DLT beams (Series 2-1)

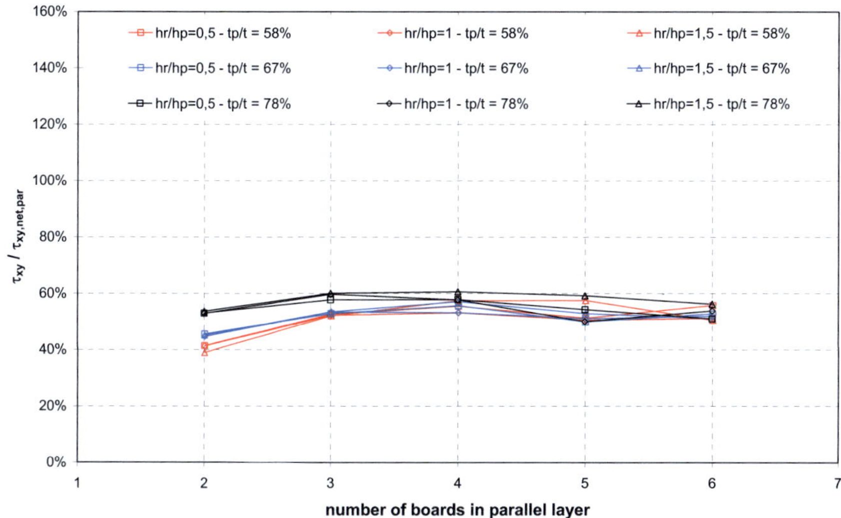

Fig. 134   Ratio between τ_{xy} and $\tau_{xy,net}$ for short 4-ply DLT beams (Series 2-2)

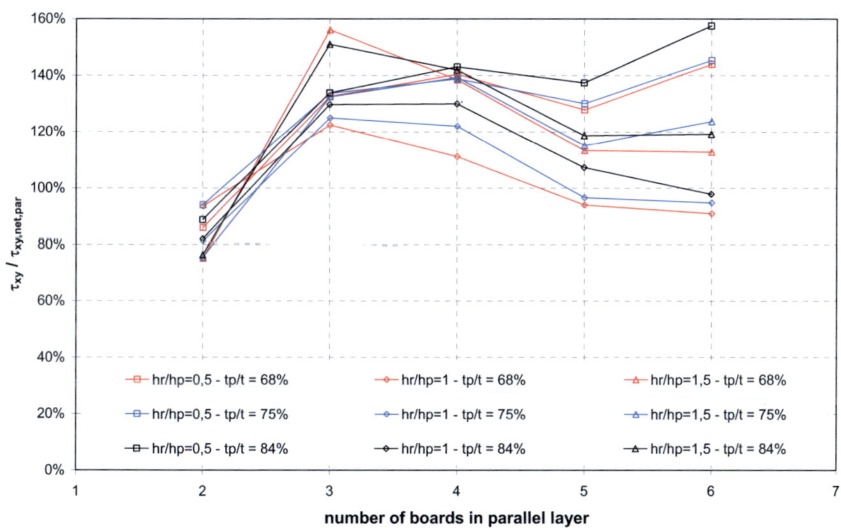

Fig. 135   Ratio between τ_{xy} and $\tau_{xy,net}$ for long 5-ply CLT beams (Series 2-1)

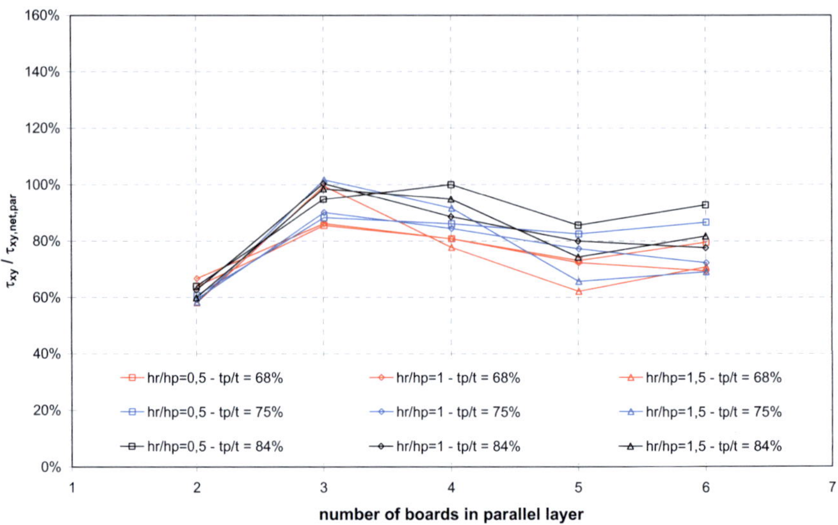

Fig. 136   Ratio between τ_{xy} and $\tau_{xy,net}$ for long 5-ply DLT beams (Series 2-3)

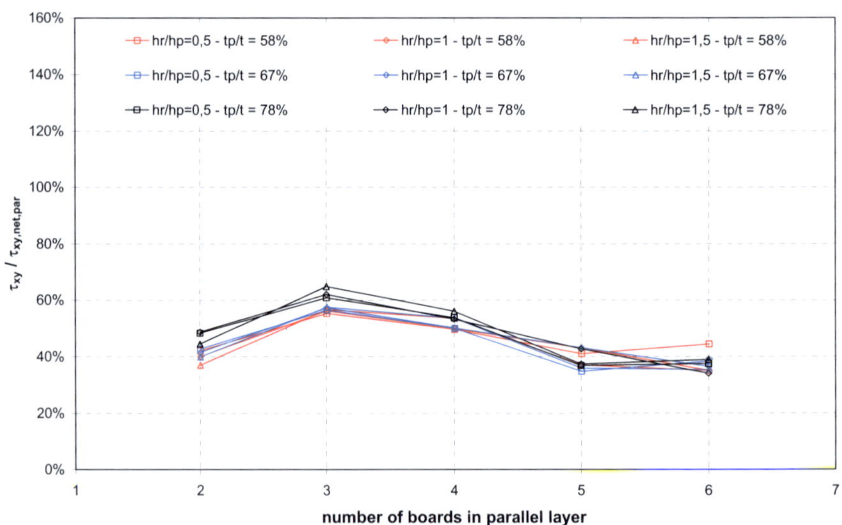

Fig. 137   Ratio between τ_{xy} and $\tau_{xy,net}$ for long 4-ply DLT beams (Series 2-4)